M Walton
Dpt. of Psychology

D1316189

Dr. CRYPTON and HIS PROBLEMS

Dr. CRYPTON and HIS PROBLEMS

Mind Benders from *Science Digest*

by Dr. CRYPTON

Illustrated by Matt Freedman

ST. MARTIN'S PRESS • New York

DR. CRYPTON AND HIS PROBLEMS. Copyright © 1981, 1982 by Science Digest, copyright © 1982 by Dr. Crypton and Matt Freedman. All rights reserved. Printed in the United States of America. No part of this book may be used or reproduced in any manner whatsoever without written permission except in the case of brief quotations embodied in critical articles or reviews. For information, address St. Martin's Press, 175 Fifth Avenue, New York, N.Y. 10010.

Design by Dennis J. Grastorf

Library of Congress Cataloging in Publication Data

Crypton, Dr.
    Dr. Crypton and his problems.

    1. Scientific recreations. 2. Mathematical recreations. 3. Physics—Popular works. I. Science digest. (Hearst Corporation: 1980) II. Title.
Q164.C78   593.73   82-5720
ISBN 0-312-21476-6   AACR2

10 9 8 7 6 5 4 3 2

Contents

To people with problems
everywhere

Acknowledgments

MUCH OF THE MATERIAL in this book appeared in my column "Puzzles, Paradoxes and Pitfalls" in *Science Digest* magazine. I want to thank the staff of *Science Digest,* particularly Scott De-Garmo, Pierce G. Fredericks, Kathy Guthmuller, Judith Stone, and Russell Zolan, for letting me unleash my wacky conundrums on their readers. I also want to express my gratitude to St. Martin's Press, particularly to Michael Denneny and Paul Dinas, for giving me this additional opportunity to play mind games with the world.

Many of the brainteasers in this book have been fine-tuned because of the letters I've received from more than ten thousand puzzle enthusiasts. I'm glad that my column has helped to bring together the problem-solving community, and I hope this book will do the same. To this end, I invite you to write me care of *Science Digest,* 888 Seventh Avenue, New York, N.Y. 10106.

Questions

PEOPLE ASK QUESTIONS, for one thing, because they want the answers and because other people have the answers. If Smith asks Jones, "What's your phone number?", Smith presumably lacks that needful bit of information while Jones has it, of course. The question elicits the transfer of knowledge.

Such a ritual, however necessary, is dullness itself. Who has ever filled out a questionnaire without finding all that listing of name and address and birthplace and college degree and all the rest of it unbearably boring?

The trouble is that you know the answers too well. There's no fun at all in an answer that's routine. For that matter, there's no fun in an answer that you simply don't know, and *know* you don't know. How do you recite the 23rd Psalm in Sanskrit? I don't know, and there's an end to it.

Suppose, though, that uncertainty rears its ugly head. How old was Queen Victoria when she died? Who won the World Series in 1938? Maybe I know—and maybe you disagree with me—and maybe the cry of "Bet! Bet!" rings out. Once it becomes a matter of financial competition, with what glee all troop off to the reference books so that there might be occasion for the passing of funds, for triumph and chagrin.

But it doesn't take filthy lucre to rouse interest. The best of all questions are those that don't require a reference book, but merely a bit of thought—something you don't actually know offhand, but can think out with a bit of mental effort. Almost everybody is amused by a question which has an answer that is not obvious, until it is explained, or worked out, or stumbled across—and which then becomes obvious.

The answer is then liable to evoke the same kind of laughter a good joke does, or the same kind of groan a bad pun does, and leaves us with an almost unbearable impulse to try it on someone else.

For instance, the greatest West European monarch of the Middle Ages was Charlemagne. He ruled over all of Latin Christendom (except for the British Isles) for over forty years, was uniformly successful in war, kept his realm in order, was crowned Emperor by the Pope, and so on.

Now here's a question! What was Charlemagne's name? I'm serious. What was his name?

It certainly wasn't Charlemagne. Neither his father, Pepin, nor his mother, Bertha, ever chucked him under his chin and said, "And how's little Charlemagne today?"

His name, in Latin, was Carolus; in German, Karl; in French and English, Charles.

By universal consent, however, his successful reign caused him to be called, in eventual hindsight, "Charles the Great"; or, in German, "Karl der Grosse"; or, in Latin, "Carolus Magnus"; or, in French, "Charles le Magne." The last version, shortened to "Charlemagne," came into universal use in French and English, but that doesn't make it his name. It's merely what he is *called*.

Well, then, why is Dr. Crypton's name? No, not what; *why?* It's not the real name of the author of this book, of course; it is a pseudonym. But why *this* psuedonym?

As it happens, the Greek word *kryptos* means "hidden." That's why a "crypt" (the Greek "k" becomes "c" in Latin and, hence, in English) is a room that is underground and, therefore, hidden. A "cryptogram" is a writing in code with a message that is, therefore, hidden. And anything that is "cryptic" has a significance that is ambiguous or uncertain and, therefore, hidden.

Naturally, Dr. Crypton's name signifies that he specializes in questions and puzzles to which the answers and solutions are not obvious; and which are therefore hidden, and, in consequence, *fun*. It is for that reason that your enjoyment of this book will be as great as Charles. (I'd ask you what I mean by that, but, having read this introduction, you already know.)

—*Isaac Asimov*

Dr. CRYPTON and HIS PROBLEMS

CHAPTER 1

An Enigmatic Evening with Microwave Cornbread

I HATE TO ENTERTAIN, but once a year, on March 7, I force my-self to make dinner for three of my co-workers at the Institute for Paradoxology. In 1982 the little dinner party included Lora G. Huss (my secretary), Ron C. Plute (the psychoanalyst), and Hector Pomic (my boss). There's no point describing what these people look like because their names are anagrams that do just that.

Each year the same thing happens. I whip up some light dish. Everyone eats it, gets smashed, and tries to humiliate his co-workers at a variety of board, card, and mind games.

This time my guests were late so I turned on the television. The anchorman was wrapping up his coverage of the day's sports: "And so it was a hitters' duel in the nine-inning game between the Pittsburgh Panthers and the Boston Bunters. The Bunters scored three runs an inning, while the Panthers tallied two runs an inning. Six batters belted home runs over the wall of the new Massachusetts stadium. Remarkably, the Bunters won without a single man ever crossing second base. Stay tuned for a breaking story: 'The Flight of the Killer Bees.'"

I turned off the set. I wish I had put it on at the beginning of the sportscast because I couldn't figure out how the Bunters had earned their runs. Maybe you can. I was also too lazy to make the simple computations necessary to determine the final score. Why don't you make the calculations for me?

The doorbell rang. I opened the front door, and there was Lora in a Möbius-strip dress. Lora has two of the best conic sections I have ever seen.

"I'm sorry I'm late, Dr. Crypton. The queue at the bank was much longer than I expected. And while I was waiting, the woman behind me tapped me on the shoulder and asked me whether I had change for a dollar. To my tremendous embarrassment, I found I had the most amount of money in coins I could possibly have without being able to give her change. Still, she didn't want to wait for the teller, and so she took a small loss by giving me the dollar for whatever change came closest to a dollar without exceeding it."

"Well, Lora, you need not apologize for your tardiness since you profited from it." I smiled and mixed her a double Scotch and soda.

"Don't be so coy, Doctor. You and I both know that all the money I have on me will be yours before the evening's over. You and your devious little bets and games!"

"Come now, Lora. That's no way to treat your host. I'm genuinely glad you made money at the bank."

"That's not all I made there. When I finally reached the teller, he was so exhausted that he misread the trust-fund check I cashed. He gave me dollars where he should have given me cents, and cents where he should have given me dollars."

"How much money was the check made out for?"

"That's none of your business. You old scoundrel! Just because you're my boss, you have no right trying to find out how much trust-fund income I get."

Lora was spunky tonight. I could see she would be a fierce opponent in the evening's games. "No, no. You misunderstand me," I told her. "I only want to know how much you profited from the teller's error."

"Oh, I can tell you that. I got ninety-nine cents more than I should have. But I tipped the teller with the change he gave me."

"How much was the tip?"

"It was equal to the amount of change I was left with after I had given the woman partial change for her dollar."

"I see." I tried to disguise my immense delight. Lora had unwittingly provided me with enough infomation to calculate her trust-fund income. Now I could determine if she was worth chasing. You know how much I long to give up my chair at the Institute and live off some woman's wealth. See if you, too, can calculate her trust-fund income.

The doorbell rang again. It was Ron and Hector. "Good to see you, gentlemen," I told them. "How are you guys?"

"We're fine. And so are you," replied the psychoanalyst.

"That's good to know," I said cheerfully. "Lora is in the living room. Why don't you join her while I prepare dinner? The alcohol is in the alcove on the left."

As they helped themselves to the spirits, I put a circular cornbread cake into the microwave oven. For the cake to taste best, it would have to bake exactly nine seconds. Unfortunately, the oven timer was broken. The only other timers I had were a seven-second hourglass and a four-second hourglass. Was I able to use them to bake the cake?

The cornbread was ready in a jiffy, and so I took it into the living room to serve to my guests. I was about to cut it into four equal pieces with two knife strokes when Hector stopped me. "Dr. Crypton," he said, "by cutting big pieces you're maximizing the possibility that someone won't finish his cake. It would be better to cut the cornbread into eight pieces. That way each of us can eat two pieces if his appetite holds up. But if someone is full after the first piece, his second slice will be available for the rest of us to eat."

"You have a good point," I said. I pictured the cake neatly divided into eight slices. I wanted them to be perfect, so I practiced by moving the knife up and down above the cake in the four places I intended to cut it.

"Dr. Crypton?" Ron said somewhat hesitantly.

"Yes?" I replied.

"It's a wasted effort for you to make four incisions. You can divide the cake into eight pieces with only three cuts."

"Thank you. You're quite right," I told him. With the tip of the knife I traced out on the top of the cake the places where I'd make the three cuts.

Lora gulped. "Don't cut it that way. You're making the pieces different sizes."

I was mortified when I realized that she was right. In a second, however, I was able to cut the cake three times and get eight equal pieces. How did I do that? And how did I intend to make the cuts before Lora had corrected me?

As it turned out, all of our ingenuity was for nothing because we each ate two slices. A few cocktails later, the four of us were eager to start playing the evening's games.

Lora demanded that Ron play cribbage with her, and Hector challenged me to a game of chess. Normally, I would not have accepted because I hate playing weakies. You sit there for twenty minutes while they study the position intensely. And then after all that time they come up with some horrible move. Moreover, it's no fun beating them because their egos aren't crushed. They're so accustomed to losing that they take it in stride. I was drunk enough, however, to accept Hector's challenge. And as a handicap I promised him that I would never put a knight on any square adjacent to his king.

After he misplayed the Fried-Liver Attack, I found myself with an overwhelming position. I could mate him on the move in several ways, but I hesitated because he is, after all, my boss. I thought it best to prolong the game, but I had difficulty finding a move consistent with the handicap that would not checkmate him. Can you?

White: Dr. Crypton　　　　　　**Black: Hector Pomic**

While Hector and I finished the game, Lora and Ron went outside on the patio to play ticktacktoe. I could overhear their conversation, and I didn't like the risqué direction it was taking.

I heard Lora ask: "Have you ever looked at Eric Partridge's dictionary of the bawdy words in Shakespeare?"

"Indeed I have, Lora. I remember being amused by Partridge's prudish definition of 'nest of spicery.'" Ron paused, and I quivered at the thought of what they might be doing during the silence. Fortunately for my delicate constitution, Ron started up again. "Partridge defined it as 'the pudend and the circumambient hair.'"

"That's very funny," Lora said. "I wouldn't expect a lexicon of lewd words to beat around the bush. Ron, since you're a psychotherapist, I thought you might be able to give me some advice. My sister is trying to decide whether she should be in analysis."

"Well, Lora. It all comes down to . . ." Ron lowered his voice. "To how often she has orgasms."

"She has them all the time, Ron."

"Hummm? And how often does she engage in sexual congress?"

"She *never* does."

"That's her problem. She's coming unscrewed."

That remark was too much. I dashed onto the patio before Lora and Ron could go any further. They were soused, and I could see at once that their ticktacktoe games were the brainchildren of intoxicated minds. For example, could the following game have ever been produced by strict adherence to the rules?

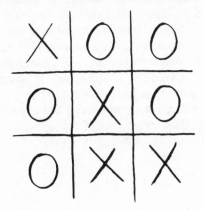

I led Lora and Ron back into the living room, where Hector proposed that we listen to music.

"Drat! The record player's broken," I told them.

"That's okay," Lora said. "I'll just play some songs on your Touch-Tone phone." She picked up the phone and pressed the following buttons (a comma indicates when she paused): 3212, 333, 222, 399, 3212, 333, 22321.

I was amazed to find that I recognized the song. I also knew the next tune she played: 0$\underline{11}$, 12369#, 955, 66512$\underline{966}$, 6#913$\underline{\#\#}$ (an underline indicates where she held a note).

See if you too can determine the songs by punching the sequences on a Touch-Tone phone. Actually, the best way to do this is to call up a friend and have him play the songs to you. If you do it yourself, you run the risk of accidentally dialing some exotic foreign land.

Now that we had degenerated to the point of making music with the telephone, I thought it best to send my guests home. You cannot imagine how pleased I was once they left, and the year's only social obligation had come to a close.

I did a two-step into my bedroom and took out a telegram I had been saving until then to read. It was from Iva, a girl I had seduced two weeks before. The telegram, which she had sent from a conference she was attending with her prominent boss, said: "My dearest Crypy, I send you now as many kisses as the largest integer known to man whose only divisors are one and itself."

I swooned with delight as I tried to imagine the magnitude of that integer, called a prime number. A mathematics journal I kept by my bed indicated that in 1975 two high-school students in Hayward, California, used a computer to discover the largest-known prime: the 6,533-digit number $2^{21,701}$ minus 1. The nice thing about Iva's message is that it will mean more every time a larger prime is found. If you happen to stumble across one, please let me know in a letter addressed to *Science Digest*.

By the way, when I was reviewing the evening's festivities in my mind, I realized that one of my guests had told an outrageous lie. Do you know what it was?

When I originally intended to cut the cake into eight pieces by making all three cuts in the top of the cake, I planned to make two vertical cuts, forming a cross, and one vertical cut, tracing a circle.

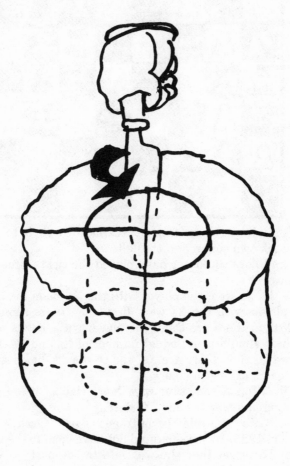

Sliced Cornbread

The only move that doesn't checkmate Black is to move the White rook two squares to the left.

Ron and Lora could not have played the ticktacktoe game by the rules.

Since the Xs won, the player with the Xs made the last move. If the player with the Xs went first, then there should be one more X than there are Os. If the player with the Os went first, then there should be an equal number of Xs and Os. But there are more Os than there are Xs, and so the doodle doesn't represent a game of ticktacktoe.

The first Touch-Tone tune was "Mary Had a Little Lamb" and the second one was the hit "Yesterday."

It was Lora who told the outrageous lie. March 7 fell on a Sunday in 1982, and so she could not have gone to the bank. She thought I'd forget that she was late to the party if she could entertain me with an engaging story. Since the business about the teller was contrived, I have no reason to believe that her trust-fund income is as low as $21.20. She might be worth chasing after all.

LETTERS

A version of "An Enigmatic Evening with Microwave Cornbread" appeared in the August 1981 issue of *Science Digest*. Many readers sent me alternative solutions to the puzzles that came up in the course of the dinner party.

David Fischer of Gahanna, Ohio, provided another explanation of how the Boston Bunters could have won a game without a single man ever crossing second base. "You proudly proclaim that the Bunters is an all-women team, proving you're not a male chauvinist (until a few paragraphs later when you declare that you want to 'live off some woman's wealth'). Isn't it also possible, however, that every player who crossed second base was a married man?"

Another correspondent, who undoubtedly failed English composition, sent me an incomprehensible letter, the farfetched thrust of which seems to be that if a base runner caught up to the man in front of him, they could cross second base together. That way it's not a single man but two men that cross the base.

Jeanne Goodman of Morgantown, West Virginia, pointed out that although $1.19 is the largest amount of money Lora could have in coins without being able to change a dollar, she need not have three quarters, four dimes and four pennies. Another possibility is that Lora could have a quarter, nine dimes and four pennies. "I knew all those years of clerking would be useful someday!"

Bill Dembowski of Johnstown, Pennsylvania, noted that fans of Susan B. Anthony and Dwight D. Eisenhower will realize that Lora could have much more than $1.19 and still not be able to change a dollar. Indeed, the amount of money in coins is limited only by the size of her pockets and handbag.

"I would also keep a close eye on your psychoanalyst," Dembowski warned, "because his name, Ron C. Plute, is an anagram not only of *corpulent* but also of *cult prone!*"

Frank M. Gregario of Wappinger Falls, New York, provided an alternative way of timing the microwave cornbread. My solution, which makes use of both hourglasses more than once, requires flipping them a total of five times. But suppose the four-second hourglass was from a shlock house and the seven-second hourglass from a chest of the personal effects of the first Cryptons to come to America. To avoid damaging the precious heirloom I would use it as little as possible. With that in mind, how would I use the two hourglasses to time the cake? Needless to say, this method of timing must require no more than five flips; under no circumstances would I let nostalgia stand in the way of efficiency.

Gregario's answer is clever because the oven is started not at the beginning of the timing sequence but in the middle. Put the cake in the microwave but do not turn it on. Start both hour-glasses. When the four-second hourglass has drained, flip it over. When the seven-second hourglass has drained, start the microwave. At that point there is one second remaining in the four-second hourglass. When the one second has elapsed, flip over the four-second hourglass, let it drain, and then flip it over again. When it is empty, the cake will have baked nine seconds and the precious hourglass will have been used only once.

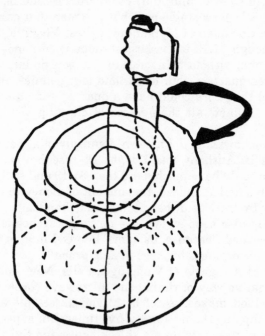

Bernard's Cornbread

The same solution was submitted by Bob Bernard of Sarasota, Florida. Bernard also pointed out that I could have saved myself work by dividing the cake into eight slices not with three cuts but two. The trick is to make one of the cuts in the form of a spiral, as the illustration indicates. This method can yield any number of pieces with only two cuts. Nevertheless, my method of cutting eight pieces at least had the virtue of ensuring that they all were the same size.

Several correspondents recognized that in the chess game the weird pawn structure could not have come about if Hector and I had been playing strictly by the rules. There is no doubt that the cocktails were responsible at some point for a cavalier interpretation of how the pawns move.

Iva's telegram means even more to me now because of letters from Ted Stockwell of Burnsville, Minnesota, and Jim Clifton of Port Richey, Florida. They alerted me to the discovery of a larger prime: the 13,395-digit number $2^{44,497}$ minus 1. Clifton also told me that $2^{23,209}$ minus 1 is prime. Keep those prime numbers coming!

CHAPTER 2

The Intragalactic Cabaret

I NEVER THOUGHT I'd do something faddish, but I finally gave in to the home-computer craze and bought a medium-size model for my upstairs bathroom. As I lay lathered in the tub, I would be able to compute such things as the odds that a satellite would fall out of the sky and crash into my house. I had asked my secretary to pick out the computer, and she was setting it up in the bathroom.

"I chose a great model for you, Dr. Crypton. The instructions claim that the bright spot on the screen of the cathode-ray tube moves faster than the speed of light."

"Lora, I hope that's not the reason you picked the model. You know the claim is misleading. According to Einstein's special theory of relativity, neither mass nor energy nor information can travel faster than light."

"Oh, Doctor. That relativity stuff was dreamed up at the turn of the century. It's outdated now. I can think of plenty of things that exceed the speed of light [186,000 miles per second]."

"You can?"

"Sure. Just imagine you have an ultrapowerful searchlight whose beam you directed at a distant cloud. The beam would illuminate a small circular patch. By jerking the searchlight you could make the bright spot sweep across the sky faster than the speed of light."

I was astounded by Lora's astute analysis. There was absolutely nothing wrong with her reasoning. The spot would indeed exceed the speed of light. Just picture it. If you started by aiming the searchlight vertically, so that the bright spot was directly above your head, then you could make the spot travel halfway across the sky in a fraction of a second just by flicking your wrist so that the searchlight became horizontal. Nevertheless, this phenomenon does not violate the theory of relativity. See if you can figure out why before you read on.

"Lora, your example is clever but deficient because it rests on a common misunderstanding of relativity. There are all sorts of phenomena taking place at a superluminal speed (that is, faster than the speed of light), but they don't violate relativity because they never involve mass, energy, or information moving that fast. The bright spot in your searchlight example is not a single physical object. Light consists of massless particles called photons, and the photons that make up the bright spot are different ones from moment to moment. This means that not one photon exceeds the speed of light. The illusion of superluminal motion comes about because your mind treats the spot as if it were a single physical object."

"I see your point, Dr. Crypton. Since the spot is not one object but a constantly changing set of photons, no mass or energy is moving faster than light. Nevertheless, the theory of relativity also maintains that information cannot be transmitted at a superluminal speed. I think that my searchlight example proves that this aspect of relativity is incorrect."

"You do?"

"Yes. Suppose two women, call them Gertrude and Florence, each sit on their own cloud some distance apart. Gertrude has a laser gun whose deadly rays travel, of course, at the speed of light. She gets her kicks by firing the gun every now and then at Florence. Because Gertrude likes Florence and means her no real harm, they devise a program of laser blasts that enables Florence to duck out of the way of the lethal rays. Their program calls for your participation. They want you to sweep the sky with your ultrapowerful searchlight so that the bright spot will move with a superluminal speed from Gertrude to Florence. Moreover, they agree that when the spot illuminates Gertrude she will fire her laser directly at Florence's head and that when the spot reaches Florence she will duck. Since the spot moves faster than the laser light, Florence is warned of the approaching blast. Is this not, I ask you, a clear example of a faster-than-light message?"

I started to laugh. I was so pleased with myself for having chosen such a clever secretary. She could certainly think on her feet. I looked forward to her handling my phone at the Institute for Paradoxology, where she'd be able to poke holes in the puzzles, riddles, and arguments that people phoned in. In spite of her brilliance, there was a flaw. Can you find it?

"Lora, your example of the laser gun is ingenious because it helps us to clarify what Einstein meant by a message: namely, the propagation of information from one point to another point. Here no information is being transmitted from Gertrude to Florence. The two women are merely actors completing their assigned parts in a play whose curtain time I control. I am the one who decides when Gertrude will fire and when Florence will duck. And according to relativity, what I do with the searchlight cannot be thought of as the transmission of a superluminal message because it is not the spontaneous transmission of information but the completion of a prearranged act."

"All right. I give up with the searchlight. But I still believe that I can think of things that exceed the speed of light."

"Okay. Try again."

"Well, consider a pair of scissors whose blades are a few miles long. As the scissors are closed, the point of intersection of the bottom of the top blade and the top of the bottom blade moves toward the tips of the blades. If you shut the scissors fairly quickly, the intersection point would ultimately move with a superluminal speed. In fact, it would move faster and faster as you continued to close them."

"Wait. Before you go any further, I want to see if what you say is true." I was able to remember enough of my high-school physics and trigonometry so that I could write an approximate equation for the motion of the intersection point. Sure enough, the intersection point would move superluminally. I told Lora to continue.

"This time, Dr. Crypton, I'm talking about a single point. This means the observation you made about the searchlight spot being not one object but a set of ever-changing objects does not apply to the scissors. So as I see it, the scissors topple relativity theory."

Once again Lora had made a subtle mistake. Can you find it?

"Actually, Lora, you're quite wrong. The examples of the scissors and the searchlight collapse for the same reason. Like the searchlight spot, the intersection point is not a material object but a geometrical construct. The electrons, protons, and neutrons that make up the intersection point vary from moment to moment so that not a single bit of matter is moving with a superluminal speed."

"You've outsmarted me again, Dr. Crypton. But I can think of something that proves that information can be transmitted faster than the speed of light. Are you familiar with the Intergalactic Cabaret?"

"I've heard of it, Lora, but I don't keep up with popular culture."

"Well, Doctor, whenever Venus eclipses Mercury a big show is held on the asteroid Nagas. The asteroid is so small, however, that the stage hands and technicians do their work from the planet Ofu a light-year away. Bucky, the lighting engineer, has to turn on the spotlight, located on Ofu, a year before the show because it takes that long for the beam to reach Nagas and illuminate the stage. Isaac, the director of the show, extends a cane from Ofu to Nagas so that he can yank lousy performers off

the stage. The last show was critically acclaimed, particularly because of the dancing bears, Ursa Major and Ursa Minor. The only disappointment was the soft-shoe routine. Neither Bucky nor Isaac could stand to watch it; each acted independently to cut short the routine. Bucky turned the spotlight off and Isaac used his cane to pull the tap dancer from the stage.

"Clearly, it is Isaac's action that had an effect. Since the cane is rigid, when the Ofu end of it is moved, the Nagas end moves too. The performer is pulled from the stage as fast as Isaac can manipulate the cane. Bucky's action, however, has no effect on the performance because it will be a year before the stage is dark. The photons already in flight between Ofu and Nagas will continue on their cheery way when the spotlight is turned off. Is this not a clear example of a message being transmitted faster than the speed of light? Doesn't the performer instantly get the message that his routine stinks?"

I heard what Lora said, but the warmth of her soft brown eyes temporarily short-circuited my rebuttal resources. When I regained my composure, I realized she had made a mistake.

"There is no such thing, Lora, as a perfectly rigid object. Although a cane may seem rigid, it is actually a conglomeration of molecules. Motion at one end does not instantaneously transmit itself to the other end. Instead, the motion is transmitted as a ripple along the cane as one molecule joggles the next one. The velocity of the ripple depends on certain properties of the cane, such as the kind of material, but in no circumstance does the velocity exceed the speed of light."

"You win. I'll stop trying to refute relativity. I was wondering, though, what it is about the theory of relativity that prohibits an object from moving faster than light."

"Well, Lora, the relativistic equations describing the motion of objects that have mass indicate that as an object is accelerated to near the speed of light, it increases in mass. Ever-increasing amounts of energy must be pumped into the object to get it to travel still faster. To force it to move at the speed of light would require an infinite amount of energy. Since there exists no infinite source of energy, no object can be accelerated to that speed."

"I understand what you're saying about the relativistic equations prohibiting the acceleration of an object to the speed of light. But why should we believe the equations? They're not laws but conventions. They're merely convenient mathematical representations of how matter and energy have behaved in the past. A bunch of variables strung together by algebraic signs cannot dictate how matter and energy will behave in the future. Any moment now, an object might be accelerated to at least the speed of light. Then we'll have to throw out your damn equations and revise our physical theory."

"No, Lora. That won't happen. The relativistic equations have successfully explained countless phenomena in physics. In fact, the equations have never failed. The scientific community accepts the equations because of their well-documented explanatory power. A particle accelerated to the speed of light could not be treated as an isolated exception. We could not react by saying, 'Oh well, relativity is not applicable here,' and go about our merry ways. The existence of such a particle would mean that the explanatory power of relativity was only an illusion. No physicist is prepared to accept that conclusion."

"Your explanation sounds reasonable enough, Dr. Crypton. I see that the equations rule out acclerating an object to a luminal or superluminal speed. But what would happen if an object moved faster than light from the moment of its creation? That would avoid the difficulty of infinite energy because the object would never have to be accelerated to that speed."

"You're right. Such hypothetical objects are called tachyons (after the Greek word *tachys,* meaning "quick"). These superluminal particles would be completely consistent with relativity theory.

"They would have many amazing properties that would distinguish them from ordinary subluminal particles. Tachyons would lose energy as their speed increased, unlike slower-than-light particles, which would gain it. Tachyons could never move slower than the speed of light, and the number of them in a region of space would depend on the nature of the observer who looked at them. Nevertheless, although they are consistent with the accepted laws of physics, there is now no observational evidence that they exist."

"I have one final question, Doctor. Should we sue the manufacturer of the computer I got you for stating that the bright spot on the cathode-ray tube can move faster than the speed of light?"

"No, Lora. The statement could be technically correct. If the spot on the tube is a projection of a beam, like the searchlight spot, then it could move with a superluminal speed. Of course, as we've seen, such motion could not be harnessed to transmit a message."

"Oh, Doctor, I've thought of something else that exceeds the speed of light."

"What is it this time?"

"The speed with which you shoot down my arguments."

"Oh, Lora."

I want to close with two problems involving velocity.

1. Suppose you put two flashlights back to back so that their beams project in opposite directions. Now look at a particular photon in the left-hand beam and a particular photon in the right-hand one. Since the two photons are moving in opposite directions (each with the speed of light), they are separating from each other with twice the speed of light. Why doesn't this phenomenon violate relativity?

2. A rocket moving with .8 the speed of light takes off from the earth for Pluto. Four minutes later a missile moving with .7 the speed of light is launched from Pluto toward the earth. The missile and the rocket eventually collide. At the time of the collision, which projectile is closer to Pluto? (If you look up the distance between the two planets in an astronomical table, use the value for their maximum separation.)

ANSWERS

1. Like the searchlight example and the scissors effect, the back-to-back flashlights do not violate the theory of relativity because no matter, energy, or information is moving faster than light. Each of the photons under scrutiny is moving with the speed of light. The distance between the diverging photons is not a physical object but a geometrical construct, and we've seen that such constructs can move superluminally.

2. The rocket and the missile are the same distance from Pluto at the time of the collision.

LETTERS

A version of "The Intragalactic Cabaret" appeared in the May 1981 issue of *Science Digest*. Hundreds of readers wrote to me and challenged Lora's discussion of the searchlight effect on different grounds from mine. They questioned the premise that the spot would appear to sweep across the sky with a superluminal velocity. Their argument goes as follows: Although the searchlight spot is made up of a different set of photons from moment to moment, each of these photons travels from the searchlight to the distant cloud not instantaneously but with the speed of light. Lora assumed that the change in the position of the searchlight leads to an instantaneous change in the searchlight spot.

Many readers saw the situation as being analogous to squirting water out of a hose at a wall. If the hose is swept across the wall, the spot of water on the wall will not move with the hose but lag behind it. The spot can move no faster than the water squirts out of the hose. By the same token, the bright spot on the distant cloud can move no faster than the photon emanates from the searchlight.

The preceding argument is deficient. Of course it is true that a change in the position of the searchlight does not translate instantaneously into a change in the searchlight spot. To be sure, the beam of light is not a rigid rod. Nevertheless, it does not follow that the searchlight spot cannot move superluminally. Consider a distant wall on which the farthest point, A, is illuminated. Suppose the distance between A and the closest point on the wall, B, is 100 light-seconds and the distance between B and the searchlight is 10 light-seconds. (A light-second is the distance light travels in one second.) Now flick the searchlight lickety-split so that it is aiming at point B. Assume the flick takes no time. (I know it takes *some* time, but as you'll soon see that does not matter to my argument.) Now, although the searchlight is aimed at B, the point will not be illuminated for 10 light-seconds because it takes that long for the photons to travel from the searchlight to B. At the end of 10 seconds B, is lit up, and the searchlight spot—in going from A to B—has covered a distance of 100 light-seconds. The spot traversed the 100 light-seconds in 10 seconds, and so it moved 10 times faster than the speed of light. In order to show that this situation does not violate relativity theory, one must fall back on the arguments I made to Lora.

The searchlight effect is not only an abstract phenomenon but a striking occurrence in cathode-ray tubes, as I pointed out.

Moreover, it may be the key to why four extragalactic radio sources observed in the past decade appear to be expanding with velocities between two and twenty times the speed of light. The sources may be emitting a beam of particles that strikes a distant dust cloud. A bright spot would be produced where the particles and the dust collide. If the beam is moving like the searchlight, the bright spot could move superluminally.

CHAPTER **3**

The Case of the Fatal Forfex

THIS CHAPTER MAY BE DISTURBING to some readers because it involves unmentionable blood and gore. I've included it, however, because it is a true incident from my life. Moreover, it shows that logical reasoning can be effective even in the most unfortunate of situations. The account that follows was taken directly from the police blotter.

AT 11:35 ON SATURDAY NIGHT DR. CRYPTON WAS TRANSLATING GREEK MYTHS INTO URDU WHEN HE WAS INTERRUPTED BY A FRANTIC CALL FROM LORA. SHE WAS AT A DINNER PARTY

AT THE HOME OF THE BALLERINA ANNA RESTIC. TRAGICALLY, A POST-PRANDIAL GAME HAD BECOME ALL TOO REAL. IDA WHIPPLE WAS LYING IN A POOL OF BLOOD, LOCKED IN A DEATH EMBRACE WITH A PAIR OF SCISSORS. DR. CRYPTON RUSHED TO THE UNFORTUNATE SCENE.

AND WHO MIGHT YOU BE?

OSWALD MOSES PH.D. I HAVE A HIGH CHAIR IN BIOCHEMISTRY AT THE UNIVERSITY. I DON'T LIKE GAMES, BUT I PLAYED ALONG. THIS ONE WAS WORSE THAN MOST. BUT I SUPPOSE MS. WHIPPLE ENJOYED IT EVEN LESS. I, TOO, HEARD THE PIERCING SCREAM.

NOW, HOW ABOUT YOU?

I'M MACY IMANNHOP. I'M A FILM EDITOR. IN THE DARKNESS MY DRESS CAUGHT ON SOMETHING AND RIPPED. I WAS ABOUT TO HEAD TO THE BATHROOM TO FIX IT WHEN THE SCREAM PIERCED THE AIR. YOU—SOB—KNOW THE REST.

EVERYONE WAS NERVOUS AND RESTLESS AS DR. CRYPTON PHONED LIEUTENANT GARTH ANDRIMIS, HEAD OF HOMICIDE.

CRYPTON DESCRIBED THE GRISLY STABBING AND IDENTIFIED THE MURDERER. CRYPTON THEN EXCUSED HIMSELF FROM THE DINNER PARTY AND RETURNED TO HIS GREEK MYTHS.
WHAT DID CRYPTON TELL THE LIEUTENANT?

8

THE RESOLUTION OF THE CASE OF THE FATAL FORFEX

Dr. Crypton knew that Robin Gyver committed the murder because the pair of scissors is left-handed and Gyver was the only southpaw at the party. The clues to the handedness of the partiers are all in the opening illustration. The watch on Gyver's right hand suggests that he's left-handed. The baseball glove on the bed belongs to Lora because she is dressed to play softball and because she told Crypton that she had just lost the city championship. The glove is for her left hand, and so she is right-handed. Anna Restic is right-handed because she is pouring a drink with her right hand. Macy Imannhop is right-handed because she is holding a cigarette in her right hand. Oswald Moses is right-handed because he is writing with his right hand. Crypton ruled out suicide because the victim is gripping the scissors with her right hand.

Macy leaped on Robin and pinned him to the ground until the lieutenant arrived.

CHAPTER 4

Palindromes: No Lemons! No Melon!

MY FAITH IN MODERN PSYCHIATRIC DIAGNOSIS was undermined when I met Martin Otto Nitram, who was confined to room 101 of the psychiatric ward in Aksal, Alaska. Otto, as he was known to his friends, had emigrated in a kayak from Finland, where he had worked for twenty-two years as a *saippuakauppias* (the Finnish word for a soap dealer). In the United States he thought it was his civic duty to be trained as a specialist in radar, but no one would hire him because he was prone to wild, unprovoked outbursts in which he would shout blasphemous nonsense. He was diagnosed as a paranoid schizophrenic with a religious mania and delusions of omnipotence. "The patient directs his rage at God and religion," his psychiatrist explained to me, "because he fears that God may be more powerful than himself."

When I met Otto, he was seated in a wheelchair, cackling maniacally and insulting the first woman: "EVE IS A SIEVE! EVE IS A SIEVE!" Otto told me that he was plagued by dreams bursting with enigmatic recurrent images: UFO tofu, spa maps, eel glee, senile felines, murdered rum, fresh serf, smart trams, a gold log, a drowsy sword, a gift fig, a navy van, an evil olive, a tar rat, and a megawatt Ottawa gem. His greatest joy in life was smoking kinnikinnik, a pungent tobacco once favored by the Indians of the Great Lakes area.

I asked him who he was, and he replied "MA IS A NUN, AS I AM." Soon he turned his wrath on the Supreme Being. He shouted "GOD, LIVE DEVIL DOG!" and "GOD, A RED NUGGET! A FAT EGG UNDER A DOG!"

A nurse appeared from nowhere with a sedative. "Thank God, I'm an atheist," she told him. "If I weren't one, I'd be offended by your profanity." She pumped him up with phenobarbital, released the parking brake on his wheelchair, and carted him off down the hall.

"MA'AM!" he shouted at her, and then he calmed down. "I don't mean any harm. It's true what I say, no matter which way you look at it." And, holy Moses, I realized what he meant. Whether you look at his blasphemies forward or backward, they read exactly the same. He was not a paranoid schizophrenic but a frustrated logologist overwhelmed by the richness of the English language.

A word, a phrase, or a sentence that reads the same forward as it does backward is called a palindrome. Many palindromes involve religion, undoubtedly because the word DEIFIED is a palindrome and because such wordplay was first uttered by Adam and Eve. Indeed, when Adam first saw Eve, he introduced himself: "MADAM, I'M ADAM." "HMMMH," Eve quietly snorted, a bit put off by the formality; after all, they were the only two people around. To make the conversation less formal she introduced herself simply as "EVE."

Thereafter the seduction proceeded smoothly, although J. A. Lindon of Weybridge, Surrey, England, claims that at a crucial moment Adam said: "NO! NOT NOW I WON'T. ON, ON ..."

(Lindon has written a lengthy palindromic dialogue between Adam and Eve. It can be found in A. Ross Eckler's *Word Recreations: Games and Diversions from "Word Ways,"* published by Dover.)

The first man and the first woman did not have the benefit of the *Hite Report,* but they somehow managed to communicate their needs. "PULL UP IF I PULL UP," Adam told Eve. The gardener in Eden, who was a bit of a keek (a peeping Tom), mumbled to himself: "EVEN I SAW MAD ADAM WAS IN EVE."

Days later Adam was in a contemplative mood, and he summed up their experience in Paradise: "EVE DAMNED EDEN, MAD EVE."

The palindrome is thought to have flourished next in the third century B.C. in the coarse satirical poems of Sotades of Crete. Sotades lived in Alexandria during the reign of Ptolemy II of Egypt. When Ptolemy married his own sister Arsinoë, Sotades mocked him—perhaps in palindromical verse—and was promptly thrown in jail. Sotades escaped to the island of Caunus. Eventually he was recaptured by Ptolemy's soldiers, who snuffed his career as a palindromist by sealing him in a lead crate and tossing it into the sea. To this day palindromes are sometimes referred to as sotadics.

The palindrome reappeared in the Middle Ages, when the Latin phrase SATOR AREPO TENET OPERA ROTAS was inscribed on amulets and charms. (My knowledge of the phrase comes chiefly from an article by Joseph T. Shipley in the *Encyclopaedia Britannica.)* The charms were laid on the stomachs of pregnant women to ward off evil spirits that might harm the fetus. What is remarkable about the Latin phrase is that the first word, *sator,* can be spelled from the first letter of each word; the second word, *arepo,* can be spelled from the second letter of each word, and so on. Indeed, the phrase can be expressed by a square whose words read the same forward, backward, downward, and upward:

SATOR
AREPO
TENET
OPERA
ROTAS

The meaning of the phrase SATOR AREPO TENET OPERA ROTAS is not completely clear. *Arepo,* which is apparently a person's name, is not Latin, although the other four words are. The phrase can be roughly translated as "Arepo the sower holds the wheels with care."

Ecclesiastical historians have argued about the origin of SATOR AREPO TENET OPERA ROTAS. Some scholars think the phrase is Christian because it is an anagram of "Ore Te, Pater. Ore Te, Pater, Sanas." ("I pray to thee, Father. Thou healest.") Moreover, twenty-one of the twenty-five letters in the phrase can be arranged to form a Christian cross.

```
                    P
                    A
                    T
                    E
                    R
            PATERNOSTER
                    O
                    S
                    T
                    E
                    R
```

Paternoster is Latin for "our father."

What about the other four letters (A, O, A, and O)? According to Shipley, the four letters can be added to the beginning and the end of the crosspieces because A, or alpha, represents the beginning and O, or omega, represents the end (Revelation 23:13: "I am Alpha and Omega, the beginning and the end").

```
            A
            P
            A
            T
            E
A PATERNOSTER O
            N
            O
            S
            T
            E
            R

            O
```

The phrase may not be Christian, however, because it has been found on a column in Pompeii. The destruction of Pompeii in A.D. 79 predates the establishment of Christianity.

Other Latin palindromes are not nearly so esoteric. In *Oddities and Curiosities of Words and Literature*, C. C. Bombaugh cites the lawyer's motto SI NUMMI IMMUNIS, which, freely translated, means "Give me my fee, I warrant you free." Particularly amusing is the sentence ACIDE ME MALO, SED NON DESOLA ME, MEDICA, which Dmitri Borgmann, in his *Language on Vacation*, translates as "Disgustingly I prefer myself; but do not leave me, healing woman."

Palindromes abound in other languages too. Borgmann gives the Italian EBRO E OTEL, MA AMLETO E ORBE! ("Othello is drunk, but Hamlet is blind!") and the Portuguese ATAI A GAIOLA, SALOIA GAIATA! ("Tie the cage, naughty rustic girl!"). The earliest known English palindrome was composed by the seventeenth-century poet John Taylor. He wrote LEWD DID I LIVE & EVIL I DID DWEL.

Many world leaders have spoken palindromically. After Napoleon Bonaparte abdicated to the Mediterranean island Elba, he was asked whether he could have invaded England. He astounded his questioner by responding in English: "ABLE WAS I ERE I SAW ELBA." On a modern note, the Ayatollah Khomeini, upon learning of the death of Mohammed Riza Pahlavi, apparently declared: "NO EVIL SHAHS LIVE ON." And President Reagan, after vicariously reviewing the mistakes of his predecessor, could have chortled, "STAR COMEDY BY DEMOCRATS." (If I may refer to Reagan as G.I. Ron, there is the extended palindrome G.I. RON SAW STAR COMEDY BY DEMOCRATS WAS NO RIG.)

The drawback of many palindromes, such as MAD ZEUS, NO LIVE DEVIL LIVED EVIL ON SUEZ DAM, and DESSERTS I DESIRE NOT, SO LONG NO LOST ONE RISE DISTRESSED, is that they are not likely to be spoken often in ordinary conversation because they express nonsense and violate cherished notions of good grammar and diction. A brilliant exception, from the mind of London's foremost palindromist, Leigh Mercer, is A MAN, A PLAN, A CANAL—PANAMA. Lindon came up with a palindromic parody of Mercer's creation: A DOG, A PANT, A PANIC IN A PATNA PAGODA.

In defense of their reversible witticisms, palindromists assert that although their sentences may seem a bit bizarre, they are typical of the thoughts that flow through our heads when we are dreaming. Can you not picture Carl Sagan tossing and turning as he ponders in his deepest sleep the likelihood of extraterrestrial life? Suddenly, through his right brain hemisphere flashes the thought RATS LIVE ON NO EVIL STAR.

Or think of the poor woman who dreams that her restless infant is knocking a sweet potato off the dinner table. "NO, SON," she cries, but it's too late. Before she looks down to see how the potato fared, she asks herself, "MAY A MOODY BABY DOOM A YAM?" Perhaps the baby's name is Eliot and the mother finds more than a squashed yam on the floor. Might she not wonder, "WAS IT ELIOT'S TOILET I SAW?"

Honeymooners have been known to mumble in their sleep "NIAGARA, O ROAR AGAIN" and "EGAD NO BONDAGE." The maid who tends to their sheets has terrible nightmares in which she shouts "EROS' EYESORE."

In a dream the pacifist might cry "SNUG & RAW WAS I ERE I SAW WAR & GUNS" and the mathematician might say "I PREFER PI," the mammalogist "KOALA OK," and the entomologist "DID I SEE BEES? I DID."

The structure DID I ——? I DID lends itself to palindromes. DID I NAG A PAGAN? I DID. DID I SLED AT ICE CITADELS? I DID. DID I KNOW A WONK? I DID. (*Wonk* is college slang for a studious bore.)

Orthodox psychoanalysts have paid attention to reversible writing because of its resemblance to dream speech and delusional thinking. Woody Allen's *Getting Even* contains a discussion between Dr. Helmholtz, the psychoanalytic pioneer who "proved that death is an acquired trait," and his disciple, Fears Hoffnung. Helmholtz recalls the intense rivalries that marked the early days of psychoanalysis: "Rank was infuriated. He complained to me that Freud was favoring Jung. Particularly in the distribution of sweets. I ignored it, as I did not particularly care for Rank since he had recently referred to my paper on 'Euphoria in Snails' as 'the zenith of mongoloid reasoning.' Years after Rank brought the incident up to me when we were motoring in the Alps. I reminded him how foolishly he had acted at the same time, and he admitted he had been under unusual stress because his first name, Otto, was spelled the same forwards or backwards and this depressed him."

In his mathematical games column in *Scientific American* Martin Gardner explains how the fact that *God* is the reverse of *dog* has figured in certain psychoanalytic interpretations of the neuroses of pious people. Gardner cites an amusing analysis by Carl Jung, described in *Freud's Contribution to Psychiatry*, of a religious man who involuntarily made a strange, spasmodic gesture with his upper arms. Jung and his co-workers believed that the tic grew out of an earlier, unhappy experience with a dog. Gardner sums up Jung's thinking: "Because of the 'dog-god' reversal, and the man's religious convictions, his unconscious had developed the gesture to symbolize a warding off of the evil 'dog-god.'"

Jung's analysis was not meant to be a gag. All kinds of strange things can happen when we're drawn inward.

CHAPTER 5

Sleight of Mind
in Katmandu

I'VE ALWAYS ENJOYED FOOLING PEOPLE with hocus-pocus, but I am such a klutz that I have to avoid tricks that involve sleight of hand. My conjuring is based on sleight of mind. Where the professional magician relies on his ability to palm a card or to deal from the bottom of the deck, I depend on my knowledge of numerical quirk. The hocus-pocus I perform is mathemagic. An underlying mathematical peculiarity is responsible for each trick.

I was first exposed to mathematical prestidigitation in Asia. Lora and I were sitting in the waiting room of the Katmandu airport when a Nepalese woman approached us. I wasn't too nervous because she didn't look like the sort who'd try to pin flowers on us. "Ah. You must be Americans. I am a doctoral candidate in psychology at the University of Darjeeling. My dissertation is on cultural differences in mathematical ability. It would aid my research greatly if I could ask each of you to solve a couple of problems."

I hate participating in experiments, and I found her request to be an imposition, but before I had a chance to suggest that she go try testing the Abominable Snowman, Lora told her that both of us would be honored to contribute to her research. Well, I thought, at least the problems shouldn't take too long to solve. After all, she was testing the director and the secretary of the Institute for Paradoxology.

"Good," the woman said. "I'm glad you're going to cooperate. I'm afraid, however, that I'm going to have to ask you, sir, to move a few seats away from your companion, so that neither of

you will be tempted to peek at the other's answers." She has some nerve, I thought, but I did as I was told.

"Now, why don't we start with you, madame. I'm particularly interested in fractions and the trouble people have manipulating them. Here is a scrap of paper on which is written an addition problem and a multiplication problem. Please solve them and write your solutions on this answer sheet." Lora looked at the problems, grinned broadly, and promptly recorded her answers. From her smile I could tell that the problems were easy.

"Now it's your turn," the woman said, as she took the problem sheet from Lora and handed it to me. They sure were easy. The scrap of paper had written on it:

$$\frac{1}{8} + \frac{6}{8} + 0 \quad \frac{6}{8} \times \frac{9}{6} \times \frac{8}{9}$$

For the answer to the addition problem I got $\frac{7}{8}$ and for the multiplication problem I got 1. I wrote them on a piece of paper. She collected the answers but left me holding the problem sheet.

"Well," she announced, "you both got the same answer—the correct one—for the multiplication problem, but you got different answers for the addition problem."

That's impossible, I thought. I couldn't have found the wrong answer and neither could Lora. When I hired her as my secretary, I made sure she could add one hundred fractions per minute.

"Sir," the woman asked, "can you check your answer to the disputed problem?" I did, and I got the same number as before. "Perhaps, then, it is the lady who has erred," she said, as she took the sheet from me and handed it to Lora. But Lora also got her previous answer.

For the next half hour, until our flight to Kuala Lumpur started boarding, the Nepalese pre-doc kept passing the problem sheet back and forth between Lora and me. Each of us solved the problem again and again, always getting our original answer. I became convinced that Lora had made an error and that she knew it. But I couldn't understand why she didn't admit to her mistake and put an end to this nonsense. Clearly Lora wasn't fooling the tester, who must have known the correct answer. Lora, of course, thought that it was I who had made the error. And she wondered what perverse reason was causing me to stick to my answer. Finally the woman told us that the difference between our answers was 577/72. What was going on?

When I realized some days later what the Nepalese woman had been up to, I decided to devote the next couple of months to collecting and devising mathemagic tricks. One thing I found is that there is much magic in the calendar. For example, surely you must have noticed that the forces of evil conspired in 1981 to make Friday fall on the thirteenth two months in a row. A friend asked me how often this happens. The question can be easily answered, although the pattern of years with back-to-back Friday the thirteenths may be more complicated than you imagine. Try finding the pattern.

A simple trick with a calendar starts with the prestidigitator handing a calendar to a spectator and turning his back on him. The spectator is told to choose any month. The magician then asks him whether the first of the month falls on a weekday or on a weekend. He is then instructed to pick any four numbers that form a two-by-two square array.

"Now," says the conjurer, "I want you to concentrate on all four numbers. I'm going to try to pick up the psi waves emanating from your mind." After a short while, the magician gets impatient. "You're not trying. I can't receive the signals when you're sitting there treating this as a joke. You're going to have to help me out. I want you to get more involved with the numbers. Try computing their sum and telling it to me."

As soon as the prestidigitator hears the sum, he says: "Good. Doing the addition got you to concentrate. I've received the waves. The smallest of the four numbers is ——."

The trick is then executed with a different month and any nine numbers that form a three-by-three array. If the conjurer is asked to repeat the tricks, they should be varied slightly to enhance the mystery. For example, when it comes time for him to name the smallest number, he can say: "Again you're not concentrating hard enough. I can't pick up all your brain waves. Wait a moment. I've detected something. It's not the smallest number. No. It's the largest number. It's ——." In the case of nine numbers the conjurer can even name the number in the center of the array. See if you can figure out how the tricks are done.

ANSWERS

I explicitly said that the Nepalese woman passed the sheet of paper between Lora and me. Therefore she could not have introduced a second sheet with different problems on it. What she did do, however, was to turn the sheet around so that Lora was reading the problems upside down. The problems appeared to Lora as:

$$\frac{6}{8} \times \frac{9}{6} \times \frac{8}{9} \qquad 0 + \frac{8}{9} + \frac{8}{1}$$

The multiplication problem is the same upside down, and so our answers agreed. The addition problem, however, is different. Lora got $\frac{80}{9}$ and I got $\frac{7}{8}$, which yields the stated difference of 577/72.

If Friday falls on the thirteenth of a given month, then under what conditions will it also fall on the thirteenth of the succeeding month? That will happen if the number of days between the thirteenths is a multiple of seven (the number of days in a week). Only between February 13 and March 13 in a nonleap year is there a multiple of seven days, namely twenty-eight days. In other words, if February 13 is a Friday in a nonleap year, then March 13 is also a Friday. That was the situation in 1981, but when will it happen again? Since there are 365 days in a nonleap year, the calendar advances by one day a year because 365 is one more than a multiple of seven. Therefore February 13, 1982, is a Saturday; February 13, 1983, is a Sunday, and February 13, 1984, is a Monday. In a leap year there are 366 days, and the calendar advances two days because 366 is two more than a multiple of seven. Consequently February 13, 1985, is a Wednesday. That makes February 13, 1986, a Thursday and February 13, 1987, a Friday. So we have six years' grace before the next double whammy.

Does this mean that back-to-back Friday the thirteenths occur every six years? Definitely not. Two factors can complicate the pattern. First, February 13 might fall on a Friday in a leap year, in which case the extra day in February puts March 13 on a Saturday, and the double dose of evil is avoided. Second, February 13 might fall on a Thursday in a leap year, in which case February 13 of the next year is a Saturday because the calendar advances two days, bypassing Friday. When these factors are taken into account, the number of years between double-dose years follows the pattern 6, 11, 11, 6, 11, 11, and so on. I have worked out the pattern in the table below where the asterisk corresponds to a leap year and the boldface type corresponds to a double-whammy year.

FEBRUARY 13

Friday 1981	**Friday 2009**
Saturday 1982	Saturday 2010
Sunday 1983	Sunday 2011
*Monday 1984	*Monday 2012
Wednesday 1985	Wednesday 2013
Thursday 1986	Thursday 2014
Friday 1987	**Friday 2015**
*Saturday 1988	*Saturday 2016
Monday 1989	Monday 2017
Tuesday 1990	Tuesday 2018
Wednesday 1991	Wednesday 2019
*Thursday 1992	*Thursday 2020
Saturday 1993	Saturday 2021
Sunday 1994	Sunday 2022
Monday 1995	Monday 2023
*Tuesday 1996	*Tuesday 2024
Thursday 1997	Thursday 2025
Friday 1998	**Friday 2026**
Saturday 1999	Saturday 2027
*Sunday 2000	*Sunday 2028
Tuesday 2001	Tuesday 2029
Wednesday 2002	Wednesday 2030
Thursday 2003	Thursday 2031
*Friday 2004	*Friday 2032
Sunday 2005	Sunday 2033
Monday 2006	Monday 2034
Tuesday 2007	Tuesday 2035
*Wednesday 2008	*Wednesday 2036

* = leap year

bold-face type = double-whammy year

In the first calendar trick, the question about the first day of the month falling on a weekday or a weekend was a diversion to throw you and the spectator off. The trick works itself because there is a regularity to the days in every month. The smallest number in a two-by-two square is always the first number. Let us call this number A. Then the square will be as follows:

$$A \qquad A+1$$
$$A+7 \qquad A+8$$

The $A + 7$ and the $A + 8$ entries follow from the fact that in a calendar there are always seven days between a day and the day below it. The sum of the four numbers is $4A + 16$. To find A, divide the sum by 4—and then subtract 4. In other words, when the magician knows the sum, he first divides it by 4 and then subtracts 4 to get the smallest number.

For nine numbers the array is this:

$$A \qquad A+1 \qquad A+2$$
$$A+7 \qquad A+8 \qquad A+9$$
$$A+14 \qquad A+15 \qquad A+16$$

The sum comes to $9A + 72$. To determine A the magician divides the sum by 9—and subtracts 8.

The largest number in the two-by-two array is $A + 8$. This means that the magician can find the largest number by adding 8 to the smallest number. Yet there's no need for the conjurer to determine A in order to find $A + 8$. By adding 8 to the formula for deriving A from the sum, we have a formula for getting $A + 8$ directly from the sum. The largest number is the sum divided by 4—plus 4.

Similarly, the largest number $(A + 16)$ in the three-by-three array is the sum divided by 9—plus 8. The number in the center of the array is the average of the nine numbers. In other words, it is the sum divided by 9. If you need to be convinced of this, then form an array in which the central number is B:

B − 8	B − 7	B − 6
B − 1	B	B + 1
B + 6	B + 7	B + 8

The sum of the nine numbers is 9B. Therefore B is the sum divided by 9.

The regularity of the days in each month makes possible all kinds of hocus-pocus. You might try coming up with some calendrical chicanery of your own. Please send me any good tricks you devise.

CHAPTER 6

Shockley: Sperm Hustler Offers Compatible Kinky Ladies Einsteinish Young

IT IS MY CONSIDERED BELIEF that nothing in the world is accidental. I am convinced that the names of people, particularly men of science, are really acronyms formed from words that describe their lives. Could there be any doubt that Freud stands for "First Revealed Erotic Universal Drives"? Darwin is clearly an acronym for "Demonstrated Aryan Race Was Initially Neanderthal," Franklin for "Found Result Any Nitwit Knows: Lightning Is Nasty!" and Shockley for "Sperm Hustler Offers Compatible Kinky Ladies Einsteinish Young." My own name presumably comes from "Devious Riddler Corrupts Responsible Youth; Paradoxes Torment Our Nation." Lora G. Huss surely stands for "Lady of Remarkable Attributes Gets Horny Under Scientific Stress." The many other acronyms sent to me by readers of *Science Digest* have confirmed my belief in hidden causality. Causal connections are everywhere if you take the time to look for them.

Carl Sagan was a popular target, particularly of readers in Canada. A. J. Sandy McKinnon of Saint John, New Brunswick, submitted "Stars and Galaxies Are Neat!" Scott Walsh of Toronto, Ontario, came up with "Carson's Astronomer Realizes Learning Showbiz Adds Gravity at NASA." Mark T. Ferris of Victoria, British Columbia, contributed "Stares at Galaxies and Nebulae."

American correspondents also poked fun at Sagan. Greg A. West of Charlotte, North Carolina, provided "*Cosmos* Astronomer Reveres Life: Sees a Googolplex Any Night." Irving S. Rosenfeld of Urbana, Illinois, submitted "Savant Astronomer

Sperm
Hustler
Offers
Compatible
Kinky
Ladies
Einsteinish
Young

Found
Result
Any
Nitwit
Knows
Lightning
Is
Nasty

Silly
Astronomer
Gawks
At
Nebulae

Demonstrated
Aryan
Race
Was
Initially
Neanderthal

Gravitates Around Networks." Jim and Lesley Wangberg of Lubbock, Texas, sent in "Sensationally Advertises Galaxies and NASA."

Unadulterated praise for Sagan came from Mrs. Corliss (Corky) Harrell of Crofton, Maryland: "Sagacious Astronomical Genius Assists NASA." My own attitude can be summed up by "Silly Astronomer Gawks at Nebulae."

Asimov also got mixed reviews. Greg A. West contributed "Authoring Stories Is My Only Vice." T. Torgerson of Arcata, California, submitted "Astronomer: Scientist; Intelligent Man of Verbiage." A correspondent in Katonah, New York, who has an illegible signature, sent me "Articulate Scientist: Ingenious Man of Vision."

*Authoring
Stories
Is
My
Only
Vice*

For Freud, Robert Phillips of Somerville, New Jersey, contributed "Sex In Guises Muddled Unconsciously Nurtures Dreams for Releasing Erotic Urges Daily." And to capture Jung's revision of Freud's theories, Wanda Lee King of Daly City, California, submitted "Justified Unity Not Gonads."

First
Revealed
Erotic
Universal
Drives

B. F. Skinner is another psychologist whose name readers saw through. Robert G. Ross of Dallas, Texas, provided "Some Kindly Intentions Nearly Negate Effective Reinforcement." K. R. Stahl of Port Sanilac, Michigan, came up with "Behaviorist Father Shoves Kid into Nurturing Nursery: Evaluates Results." (I would have said "Eerie Results.") Rita D. Elfering of

Behaviorist
Father

Stuck
Kid
In
Nerve
Numbing
Experimental
Room

Freeport, Minnesota, submitted "Behaviorist Father Stuck Kid in Nerve-Numbing Experimental Room."

Several readers deciphered the acronyms of inventors. For Bell, Linda Zimmermann of Nanuet, New York, submitted "Bequeathed Electrical Listening Legacy." And Land was immortalized as "Likes Automatic Negative Development" by Robin G. Theron of Saskatchewan, Canada.

Paradoxologists were also immortalized. For Rubik, Laura Conright of Shoreview, Minnesota, submitted "Renders Unsurpassed Bafflement in Kids." And for Martin Gardner, Robert Phillips of Somerville, New Jersey, came up with "Games and Riddles Done Numerically Entertain Readers."

Two readers had a field day with my name. The more charitable version, "Delightfully Refreshing Cleverly Resourceful Young Puzzler That Often Nurtures," came from Arsen Terjimanian of Troy, Michigan. A correspondent from north of the border, Vladimar Suna of Queen Charlotte Islands, suggested "Clever Ridiculous Yankee; Perhaps Totally Off Nut."

For another doctor, Dr. Frankenstein, Mr. and Mrs. Marquis of Yarmouth, Maine, submitted "Deranged Rascal Formed Radical Abnormal Namesake Knowing Eccentric Novelty Soon to Expire in Notoriety."

Clever
Ridiculous
Yankee
Perhaps
Totally
Off
Nut

E. F. Welsh of Pittsburgh, Pennsylvania, provided five clever acronyms. His submissions were for Newton ("Noticed Edible Weight Thumping on Noggin"), Goddard ("Got Off Doing Dreamy Aerospace Rocket Development"), Edison ("Eliminated Darkness in Spite of Night"), Salk ("Shot All Little Kids"), and Fibonacci ("Found Iterations Becoming Other Numbers, Adding Cumulative Counts Ingeniously").

Native
Englishman
With
Terrible
Orchard
Nightmare

Got
Off
Doing
Dreamy
Aerospace
Rocket
Development

Shot
All
Little
Kids

The apple story and *N* for "Noggin" figured in many readers' submissions. Darrell Roth of Sparks, Nevada, contributed "I Suspect Apples Are Common Nutrients Except When Tossed on Noggin." Scott Walsh provided "Injured Scientist Advances Apple Concept. Now Empiricists Weigh Theories on Noggins." And George Blahon, Jr., of Pensacola, Florida, submitted "Never Expected Wondrous Thump on Noggin." Carol McGee of West Point, New York, cleverly avoided "Noggin," but expressed the same idea: "Native Englishman with Terrible Orchard Nightmare."

For Descartes, Muriel Deitch of Smithtown, New York, submitted "Declared *Ergo Sum Cogito* (Am Reckoning Therefore Existing, See?)."

For Curie, Gary Shapiro of Norman, Oklahoma, provided "Clumsily Unraveled Radium's Interesting Emissions."

Keith Ringkamp of Lewisburg, Pennsylvania, submitted for Pasteur "Pasteurized and Sterilized the Ejected Udder Residues." I think it is unfair to use "pasteurized" to explicate the meaning of "Pasteur" because "pasteurized" comes from "Pasteur," not the other way around. To be sure, the wolf is not called a wolf because it wolfs down food or the fish a fish because it is fishy. Richard Ives of Tiburon, California, avoided this trap with "Persnickety About Sterility; Tried Efficient, Unusual Remedies."

For Leakey, Debra Grusley of Berwick, Pennsylvania, sent in "Let's Excavate and Know Everyman's Youth."

Please keep the acronyms coming to Dr. Crypton, *Science Digest,* 888 Seventh Avenue, New York, N.Y. 10106. I would love to receive an acronym for Matt Freedman, the illustrator of this book.

CHAPTER 7

Zeno of Elea: The First Paradoxologist

THE FATHER OF THE PARADOX is Zeno, a pre-Socratic philosopher who lived in the southern Italian city of Elea in the fifth century B.C. He is remembered chiefly for his conundrum of Achilles and The Tortoise, but he articulated some forty other paradoxes in the course of defending the extreme monistic world view of his intimate friend and teacher, Parmenides of Elea.

Little is known about Zeno's life. Plato described him as tall, handsome, and politically active. There is some evidence he was tortured to death for revolting against the tyrant Nearchus, who ruled Elea.

More is known about Zeno's philosophical ideas, although little of his own writing has survived. Our knowledge of his ideas comes from the work of Aristotle, who lived some one hundred years after him, and the work of Simplicius, who lived one thousand years after Zeno. Aristotle considered Zeno to be the inventor of dialectic: the method of argument in which one starts with the opponent's most basic premises, even though one might think them false, and shows that they lead to unacceptable consequences. The dialectic method was subsequently mastered by the Sophists.

Zeno employed dialectic against the belief that there is more than one object in the world and that such objects can move and change. It is no surprise that this belief was held by virtually all philosophers except Zeno, Parmenides, and a few of their cronies. Parmenides, who expressed his philosophy in the poem *On Nature,* ascribed belief in the multiplicity of objects, their motion and their change to sensory illusion. Although our senses may tell us that there is a plurality of objects existing in space and time, there is in reality only one object.

Whereas Parmenides argued directly for the existence of an eternal, indivisible One, Zeno argued indirectly. In dialectic fashion he set out to demonstrate that the popular belief in the existence of the Many logically entails further beliefs anathema to proponents of the Many.

Zeno showed, for example, that belief in the Many led to the self-contradiction that the Many is both finite and infinite in number. The Many, he argued, must be numerically finite, because there are as many things as there are, neither more nor fewer. The Many must also be numerically infinite, however, because for any two objects to be truly two rather than one, there must be a third object between them (all pre-Socratic philosophers ruled out the possibility of empty space); but for the third object to be distinct from the other two requires two more objects (a fourth object between the first and the third, and a fifth between the second and the third), and so on ad infinitum.

Zeno advanced four paradoxes to demonstrate that motion is

only an illusion. Two of them, namely The Achilles and The Dichotomy, are directed against the idea that motion is smooth and continuous. The other two paradoxes, The Arrow and The Stadium, impugned the idea that all motion is a succession of small, indivisible jerks, like the motion of film through a projector.

According to Aristotle's reconstruction of The Achilles paradox, in any race in which the fastest object has a late start, "the slowest moving object cannot be overtaken by the fastest since the pursuer must first arrive at the point from which the pursued started so that necessarily the slower one is always ahead." In contemporary philosophical literature the paradox is usually fleshed out with numbers. Achilles, who runs 10 times as fast as the tortoise, gives the slowpoke a head start of 1,000 feet. To overtake the tortoise Achilles must first get to where the tortoise started. Nevertheless, by the time he sprints 1,000 feet and reaches the tortoise's starting point, the sly reptile has crept forward 100 feet (a tenth the distance Achilles ran). By the time Achilles covers this 100 feet, the tortoise has sneaked forward another 10 feet. Achilles runs the 10 feet, but in the meantime the tortoise has slinked forward one foot. Now a one-foot run puts Achilles a tenth of a foot from the tortoise; and then a run of a tenth of a foot puts him a hundredth of a foot from the tortoise. The outcome is that Achilles can never reach the tortoise although he can get as close as his Greek warrior heart desires.

The paradox is not deactivated by pointing out that if such a race actually occurred, you would see Achilles effortlessly overtake the tortoise. Zeno was not a fool. He would not deny that if you watched the race, you would perceive Achilles to be the victor. What he would deny is the accuracy of your perception. He thought his analysis demonstrated that, in spite of your perception, it is the tortoise that is the victor.

Can you think of a way around the paradox? Alan R. White argued in the philosophical quarterly *Mind* that Achilles' heel is his aiming at the Tortoise's present position rather than at one of the future positions. White drives his point home with a conundrum of his own, Achilles At The Shooting Gallery. "The shade of Achilles," White wrote, "has given up his pursuit of the shade of the tortoise and has decided to try his luck at the shoot-

Achilles at the Shooting Gallery

ing gallery. There the targets move from left to right across his line of vision at a slow but constant speed." The shade of Zeno advises Achilles to aim at the target if he wants to hit it. Achilles follows the advice but finds that his bullet zips past the left side of the target. The Greek warrior tries faster and faster guns, and although the bullets come closer to the moving target, they still whiz by it on the left. Achilles' self-image has suffered. He reluctantly concludes that he can no more shoot a moving target than catch a creeping tortoise.

White introduces the shade of Socrates to expose the error of Achilles' ways. "Zeno has misled you," Socrates declares. "If the target moves at all, then no matter how fast your bullets and how slow the target, a shot aimed at its present position will land to the left because the target will have moved, however slightly, to the right during the time it takes your bullet to travel. You should aim for the position in which the target will be by the time your bullet has arrived; this is easily calculable from the speeds of the target and your bullet."

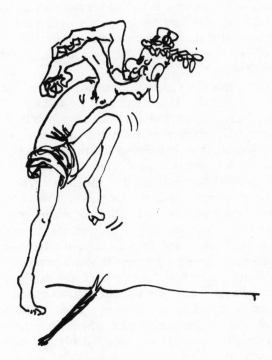

Dichotomy

Zeno digests Socrates' logic and cleverly demonstrates his newfound marksmanship by aiming at a calculated future position of the target. Of course, Socrates' logic applies equally well to the race with the tortoise. Zeno now realizes that he can catch the plodding reptile by aiming for a calculated future position.

Whatever the merits of White's analysis, it does not dispose of The Dichotomy paradox, Zeno's other conundrum of continuous motion. The Dichotomy paradox denies the existence of any motion, according to Aristotle, "on the ground that that which is in motion must arrive at the halfway stage before it arrives at the goal." In other words, for a man to walk any distance whatever, he must first walk half that distance. But for him to reach the halfway point, he must first walk halfway to the halfway point, and so on ad infinitum. In short, a man can traverse a given finite distance only if he can traverse an infinite series of finite distances. Needless to say, a man cannot traverse an infinite series of distances in a finite amount of time. Since this analysis applies to any distance, however small, no motion whatever is possible.

Some modern commentators have taken The Dichotomy paradox too lightly. They dismiss it by telling a story: When Zeno was lecturing on the paradox at a public forum, the famous skeptic Diogenes rose from the audience, announced that he would walk from his seat to where Zeno was and proceeded to do just that. (I will ignore the fact that the story cannot possibly be true—although it is often billed as such—because Zeno and Diogenes were not contemporaries.) I want to emphasize again that Zeno would not deny that the audience perceived Diogenes walk from his seat to the podium. Zeno would assert that his own analysis, based on principles of pure reason, demonstrates that the audience's perception was incorrect.

Modern philosophy has pinned too many hopes on the wise guy who emerges from the anonymous crowd at a public lecture in order to refute the lecturer's most cherished principle. It is said that at one such forum where an articulate solipsist was making the case that no one but himself exists, a mischievous empiricist bolted from his seat to the podium and pinched the solipsist. The only story in this regard that is known to be true concerns the Cambridge logician G. E. Moore. Moore was sharing with his students what he thought was a brilliant insight into the asymmetry of language: a double negative (for example, "I can't not go") is always a positive but a double positive is never a negative. A student in the back piped up: "Yeah. Yeah."

Aristotle dismissed The Dichotomy paradox because it treats space differently from the way it treats time. Do you see what he was getting at? The paradox assumes that space is infinitely divisible but time is not. Aristotle thought that if it makes sense to describe a finite length as made up of an infinite number of points, then it surely also makes sense to describe a finite extent of time as made up of an infinite number of moments. Now, if the points are paired off with the moments so that a man passes one of the points each moment, he can transverse a finite length in a finite extent of time. As the mathematician Morris Kline put it, Aristotle introduced "two senses in which a thing may be infinite: in divisibility or in extent. In a finite time one can come into contact with things infinite in respect to divisibility, for in this sense time is also infinite; and so a finite extent of time can suffice to cover a finite length."

The Arrow and The Stadium paradoxes are directed against the view that space and time, like people and pennies, come in discrete units. On this view there is an indivisible unit of time (the moment) and an indivisible unit of space (the point). In other words, any finite length consists of a finite number of points and any finite extent of time consists of a finite number of moments.

The Arrow paradox, which is the weakest of the four conundrums of motion, addresses the moments of time. Zeno argued that at each moment of its flight a flying arrow must be at rest because nothing can change position without the passage of time. Yet what is true of each moment of the time the arrow is in flight must be true of the flight time as a whole. Therefore, during the entire flight the arrow is not moving but stationary. An Aristotelian might reply to Zeno that each moment of time should correspond not to a position of the arrow but to the passage from one position to the next.

The Stadium paradox is much more compelling. It cuts to the heart of the assumption of an indivisible unit of time. Suppose there are three parallel rows of soldiers, A, B, and C, and that the soldiers in each row are directly adjacent. Now imagine that

The Stadium

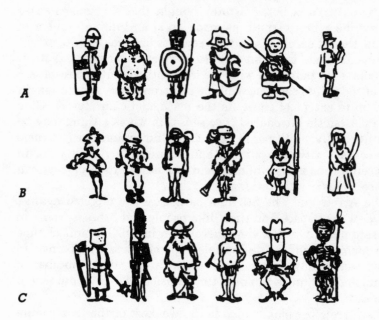

in the smallest indivisible unit of time, row A remains stationary, row B moves one position to the right, and row C moves one position to the left. The result is that relative to row B, row C is two positions to the left. At some earlier point row C must have been one position to the left of row B. Such a point, however, corresponds to an extent of time that is smaller than the smallest indivisible unit of time.

The Stadium paradox is quite clever and difficult to dispose of.

The Stadium

It is no wonder that many modern thinkers do not believe in an indivisible unit of time.

Of Zeno's paradoxes, my favorite is one that has been derided by modern commentators, The Millet Grain paradox. It can be expressed simply: How does a bushel of falling millet grain make a sound if a single falling millet grain makes no sound? After all, the bushel is no more than a collection of single grains.

Modern commentators dismiss the paradox because they know there is a threshold of sound below which the human ear cannot hear. This dismissal, however, plays directly into Zeno's overall world view. Zeno believed that we cannot trust our senses. To say, then, that a falling grain makes a sound inaudible to our ears is to concede that we cannot trust our sense of hearing. And if we cannot rely on our ears, why should we depend on our eyes when it comes to evaluating the motion of an object or the number of objects in existence?

CHAPTER 8

The Well-Hung Natives and Twenty-three Other Conundrums

WHEN I WAS A STUDENT in paradoxology at the University of Spirit Lake in Washington, my professors told me that my mission in life was to see through the veneer of ambiguity, enigma, and sophistry in life in order to discern the ultimate truths of logic, language, mathematics, science, religion, existence, and Grant's tomb. To this end, I have traveled around the world many times collecting serious and whimsical puzzles. I hope you'll send me your problems so that I can add them to my collection. Here are twenty-four of my best pitfalls. The answers, along with the comments of *Science Digest* readers, are at the end of the chapter.

The Well-Hung Natives

My pleasure boat once ran aground on the Island of Noose, which is inhabited by primitive people who follow two strange customs. First, if a wife cheats on her husband, she tells everyone but him. And second, if the husband finds out that his wife has cheated, he hangs her at midnight in the public square, leaving her body dangling.

A friendly missionary, who had been teaching these primitives Western ways, was called back to the mainland. Before he departed, he summoned the natives together and told them: "Try to remember everything I've taught you. Go easy on the firewater, genuflect often, and remember the preferred position. And I must tell you, there's been cheating on this island." The missionary set sail before the heathens had a chance to question him. Now I want you to tell me what is going to happen on the

island. This is a tough puzzle. Although the solution covers every possibility, it is quite elegant. The solution should indicate how many natives will be hanged and when they will be hanged. Good luck.

Crocked Crypton

A charming but provocative woman, Sonya Austin, told me I couldn't win her heart unless I downed one bottle of wine and twelve bottles of beer. "That will show me," she said with a malicious grin, "whether in spite of what you have written about triskaidekaphobia, you are actually afraid of the number 13." I was sitting on the floor of her Park Slope apartment, and she arranged the thirteen bottles in a circle around me. "Since you call yourself a paradoxologist, I want to make this more than an exercise in debauchery. You may start with any beer you like, but thereafter you must drink every thirteenth bottle, so that you will be fighting the dark forces every step of the way. The last bottle you chug must be the wine. It's up to you whether to move clockwise or counterclockwise around the circle. But whatever direction you start in, you must stick with it for the entire drinking spree."

Which bottle did I drink first?

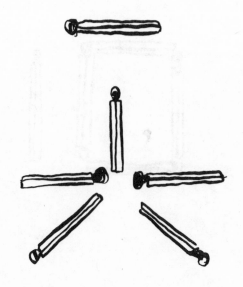

Matching Wits

The energy crunch has interfered with the patterns I make with match sticks. I used to have no qualms about arranging twelve matches to form four equilateral triangles (an equilateral triangle being one in which all three sides have the same length). But now that the price of energy has doubled, I try to make the four triangles with half the number of matches. Is that possible?

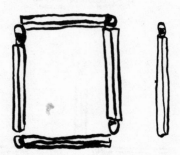

The Lighter Side

An atheist acquaintance who has a weakness for geometrical teasers asked me whether I could make a Christian cross with five matches. I paused to picture the cross on the lawn of my hometown church. "Sure," I told him. "The stem is longer than the crosspiece, so I would use three matches for the stem and one match for each half of the crossbar." He told me that my solution was correct but uninspired. Can you think of an inspired way to make a Christian cross with five matches?

No Great Shakes

Lora and I attended a party hosted by the SHARP JOKER (a fellow paradoxologist who goes by this anagram of his name) and his girl friend. There were five couples at the party, counting the host and hostess. The partiers exchanged pleasantries and shook hands. No one, of course, shook hands with himself or his mate. The Sharp Joker's friends, including his hostess girl friend, were so concerned with preserving their individuality that each one maneuvered to shake hands with a different number of people (one friend shaking zero hands, another friend shaking one hand and so on, up to eight hands). But to everyone's disappointment the Sharp Joker ended up shaking the same number of hands as one of his friends.

How many hands did the hostess shake?

A Huemuris Question

I want to share with you the celebrated puzzle of the hunted bear, which has done its part to enliven dull cocktail parties but which I have never seen solved correctly. A man with a shotgun stands due south of a bear. The hunter is a real sport, and so he goes for a little stroll to give the bear a chance to get away. He walks one hundred yards due east, aims his gun due north and kills the bear, which didn't budge a millimeter during his stroll. What color is the bear?

	CHURCHILL	HITLER	ROOSEVELT	MUSSOLINI	STALIN
YEAR OF BIRTH	1874	1889	1882	1883	1879
AGE IN 1944	70	55	62	61	65
YEAR TOOK OFFICE	1940	1933	1933	1922	1924
YEARS IN OFFICE	4	11	11	22	20
SUM	3,888	3,888	3,888	3,888	3,888

Summary Judgment

Readers of my chapter "Sleight of Mind in Katmandu" will be amused by a numerological "coincidence" I came across in Keith Ellis's *Numberpower* (St. Martin's Press, 1978). Ellis reports that in 1944 someone discovered a pattern in the lives of five world leaders: Winston Churchill, Adolf Hitler, Benito Mussolini, Franklin D. Roosevelt, and Joseph Stalin. Without exception, for each leader the sum of four numbers—the year he was born, his age, the year he took office and the number of years he had been in office—came to 3,888. What do you make of this?

Memory Lane

Hubert and Samantha spent Veterans Day weekend in a cabin in the Catskills. They were huddled in front of a fire—Samantha boning up on her math for the Graduate Record Exam and Hubert eager to have fun.

"I'll never remember which trigonometric function is which," Samantha moaned.

"Don't worry, dear." Hubert ran his hand up her leg. "I want you to keep in mind that sex on holidays can advance happiness to outrageous amplitudes."

"You're a real help, Hubert. There's so much to learn. Given a circle's radius I'm expected to compute the area of the circle to four decimal places. That's tough. I'm going to have to memorize the decimal expansion of π to four decimal places. Do you know π?"

"Yes, I know a number," he said, his eyes twinkling. "Remember that."

"You really are extraordinarily helpful. I think I'll ace the exam."

Is this a textbook example of a failure to communicate?

Backwald Tlain

On the bullet train from Tokyo to Sendai I had a long conversation with my Japanese seatmate. At one point I said: "It's not only your cars and televisions that are superior but it's your trains too."

"Quite light," he replied. "While part of your tlain is moving forward with a speed of forty miles per hour, another part is moving backwald with a speed of four miles per hour. Vely inefficient. Vely inefficient."

Was he deranged?

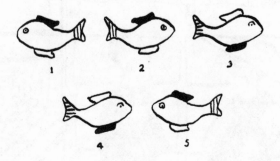

Lake Chargoggagoggmanchauggagoggchau-bunagungamaugg

On a balmy day I took Lora fishing in south-central Massa-chusetts on Lake Chargoggagoggmanchauggagoggchaubuna-gungamaugg (an Indian name meaning "You fish on your side, I fish on my side; nobody shall fish in the middle"). Lora liked putting the squirming bait on the hook, but she had no patience for casting and reeling in the line. By nightfall I had caught five fish and Lora had hooked none. What she lacked as an angler she more than made up for as an ichthyologist. "It is amazing," Lora observed, "that one of the fish does not belong with the other four." It took me a moment before I realized what she meant. Do you see?

How to Spot a Good Secretary

When the Sharp Joker wanted to hire a secretary, he put an ad in the *Sphinx*. In response, thirty-six people submitted their résumés. A preliminary screening reduced the pool of applicants to five. The Sharp Joker couldn't make up his mind, and so he summoned the finalists and had them sit in a circle.

"All of you have impressive credentials," he told them, "so we're going to play a little game, the winner of which will be given the job. The game is simple. I'm going to paste either a white dot or a black dot on the forehead of each of you. No one will be able to see the color of his own dot, but eveyone else's dot will be visible. If you see three or more black dots, I want you to stand up. The winner is the first person who figures out the color of his own dot. I want whoever figures out the color to raise his hand."

The Sharp Joker pasted the dots to the contestants. No sooner had he done so than all five stood up. After a few minutes one of the players raised his hand. Assuming everyone played by the rules, what color was the winner's dot? Also, was the game a fair one?

Deplaning with the
Motion-Discomfort Container

Besides me, Lora's favorite columnist is the crotchety William Safire, who monitors the nuances of contemporary American English in the *The New York Times Magazine*. As a professional word watcher, Safire must subject his reactions to vogue words to critical self-analysis. In his campaign for precision, discipline, and elegance in language, he doesn't want to be a fuddy-duddy, fighting single-handedly to uphold the Queen's English. On the other hand, he doesn't want to be a marshmallow and give in to trendy usage that is fuzzy or barbaric.

Lora thought Safire was at his best in a column on airline-ese. Many people are nervous about flying, and so pilots have pulled a kind of Eliza Doolittle in reverse. They have shed their educated English for what Safire describes as a highly affected "awshucks, flyin-is-jes'-a-piece-a-cake" tone of voice, which is supposed to convey that riding in a plane is like riding in a horse-drawn buggy. Airline-ese, Safire asserts, is designed to be soothing. Safety belts are now called seat belts in order not to draw attention to their function. Flight attendants refer to a doughnut-shaped hard candy not as a LifeSaver but as a mint; "they don't want anyone reaching for a piece of candy to get the notion that the pilot is preparing to ditch."

I, too, am amused by the airline industry's language of reassurance. When two planes narrowly avoid colliding in midair, the incident is called a *near miss* by airline functionaries and newspaper headline writers. A near miss, however, is literally a collision. The incident should be called a *near collision,* but the expression is much more disturbing.

Safire also airs euphemisms from the language of the flight attendant. Ask "for a barf bag and see what a look you get—not for being nauseated, but for being so uncouth not to request a motion-discomfort container." And to let you know the traumatic trip is finally over, the airline industry has promoted the hideous verb *deplane,* when the simple *exit* would have been quite sufficient. "Passengers will now *deplane* by the rear door."

I have noticed to my chagrin that the jargon of other forms of transportation also includes tongue-tripping verbs constructed from nouns. Can you think of any examples?

Sesquipedalian Fun

I wanted to rejoin the ranks of the gynotikolobomassophiles, so on the advice of my nolimetangeretarian doctor I took time off from my philosophicopsychological work on the pseudorhombicuboctahedron in order to recover from the nosocomephrenia that accompanied my hepaticocholangiocholecystenterostomy and jejunojejunostomy. Lora and I drove to Massachusetts and rented a cabin on Lake Chargoggagoggmanchauggagoggchaubunagungamaugg. The preantepenultimate fish I caught was an inedible humuhumunukunukuapuaa, so we ate lopadotemachoselachogaleokranioleipsanodrimhypotrimmatosilphioparaomelitokatakechymenokichlepikossyphophattoperisteralektryonoptekephalliokigklopeleiolaogoiosiraiobaphetraganopterygon. On the way home I proposed we see which one of us could think of the longest word. At first Lora dismissed my hippopotomonstrosesquipedelian game as bunk, but after I chided her for floccinaucinihilipilification, she agreed honorificabilitudinitatibus to play.

I ventured the obvious antidisestablishmentarianism, weighing in at twenty-eight letters. Lora replied with the forty-five-letter miner's disease, pneumonoultramicroscopicsilicovolcanokoniosis, which has been rescued from obscurity by word buffs. I came back with a whopper of a word that ended the discussion. Can you, too, do better than Lora's forty-five-letter word? Also, can you translate the first paragraph of this puzzle?

A Grave Pattern

On April 1, 1981, two days after Ronald Reagan was shot, *The New York Times* ran an article by science reporter Jane E. Brody. It began "A national emphasis on civil liberties for the mentally ill, the lack of cultural restraints upon expressions of hostility, and the ready availability of pistols are combining to make assassination an increasingly common American event, according to psychiatrists who have examined the problem.

"Assassination and assassination attempts are more common here than in any other country, experts on violence maintain. In fact, assassination is the leading cause of death of American Presidents in office."

The last sentence started me thinking about the presidents who died in office. Who were they?

Riding Between the Lines

A five-mile-long army marches in a straight line at a constant speed. A mounted courier named Dan leaves the rear of the army and heads for the front at a constant speed. When Dan reaches the front, he immediately turns around and heads at his original speed toward the rear. When he reaches the rear, Dan is where the front of the army was when he began. How far did Dan ride?

Dead Precedents

A classic storyline of brain teasers involves an allegedly smart boy who turns out to be an ignoramus. One old chestnut goes as follows: A man and a wife who have been unable to conceive a child purchase one from an unscrupulous lawyer. The lawyer promises them that the boy is bright: "He remembers anything he sees." The boy's foster parents dream about him going to Harvard someday, and so they get him a private tutor and inundate

him with high-powered history books, science texts, and famous
novels. The boy seems to like the books because he goes through
them every day. Nevertheless, he cannot remember anything in
them. Did the lawyer lie to the man and his wife?

The drawback of this teaser, and of others in the same genre,
is that the solution is easy. Consequently, I was quite delighted
when Michael Steuben sent me a more challenging problem
about an apparently ignorant boy who turns out to be smart.

Mrs. Smith was tutoring her adopted Korean son in American
history. "This is a list of our presidents in order," she said to her
son. "I remember reading that three of our first five presidents
died on the Fourth of July, our nation's birthday, although I
don't remember which presidents they were."

"James Monroe was one," replied her son.

Assuming that her son was ignorant of all American history

and politics, how did he know such an obscure fact as the date of death of a president?

Underground Expression

If you're like me, you probably don't want to confront again the kind of problem with which the school psychologist assessed your intelligence. I'm thinking of numerical sequence problems, in which you are asked to provide the next number in a string of numbers. Nevertheless, such problems need not be pedantic. Consider this charming, albeit provincial, problem sent to me by Bruce Pandolfini. What is the next number in the sequence 14, 42, 59, 86?

Math jocks should try to find the next number in the sequence 1000, 22, 20, 13. And card sharks should try to figure out the card that comes after the following four cards.

A Relatively Niece Photo to Make You Cry Uncle (Aunt You Glad I Didn't Work in Cousin Too?)

A fellow enigmatologist, Howard Sundwall of Staten Island, New York, and the American Mensa Society, described to me a recent conversation he had had with a friend:

As the friend pointed to a photo in his family album, he said: "Brothers and sisters I have none. But this man's father is my father's son."

"That's easy," replied Howard. "It's your own son. That old saw goes back to Sam Loyd [the great nineteenth-century puzzle designer] at least."

"Oh, a smart guy, eh? Then how about the fellow in this picture? He's the only son of the only brother of the only aunt of my father's only niece."

Howard thought for a few minutes. "It must be you. A picture taken many years ago, I see."

"Hey, take a look at the girl in this picture. Pretty, isn't she? She's the only niece of the father of the only cousin of the uncle of the only grandson of the brother of my cousin's mother."

Howard was stumped. He wrote me to see if I could identify the girl. Can you?

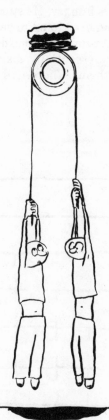

Biff and Buff

At the University of Spirit Lake, where I received my Ph.D. in paradoxology, the undergraduates were offered a smorgasbord of courses with catchy nicknames such as Spots and Dots, Nuts and Sluts, Clapping for Credit, and Physics for Poets. The nicknames had been coined by clever profs eager to boost enrollment in their courses: the history of modern art, abnormal psychology, music appreciation, and introductory physics for the mathematically deficient. I taught a section of Physics for Poets. To hold the interest of students who doubted the relevance of science, I tried to relate the principles of physics to everyday situations.

For example, consider two boys of the same mass, Biff and Buff, who are each holding the end of a rope that passes over a pulley. Above the pulley is a chocolate cake.

Buff is hungry. He summons all his strength and starts climbing hand over hand toward the cake. Biff is lazy. He just hangs there. Assume the rope does not stretch, that the pulley and the rope are essentially massless and that there is no friction between them. Where will Biff be when Buff reaches the cake?

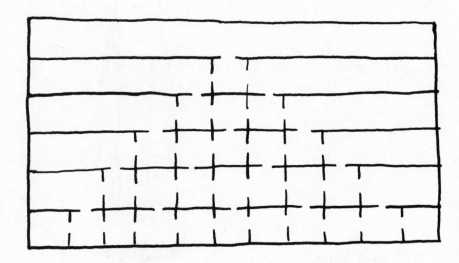

Feat of Clay

In Europe I called on Marc Swenice, the famous aesthete whose name is an anagram of the kind of art he collects. He graciously offered to let me examine his collection, which he keeps in a peculiar building with thirty-six rooms. Mr. Swenice wanted to oblige my preoccupation with puzzles, and so he proposed that I not visit any room more than once during my trip through the gallery. He said I could start my tour in any room I wanted, and he challenged me to visit a maximum number of rooms. (The floor plan of the gallery is shown.) How many rooms did I visit? Simply entering a room counts as a visit. Try to solve the problem not through the use of trial and error but by logic.

Lincoln's Downfall

At an open house at the Institute for Paradoxology, an obnoxious man was trying to steal the show by rattling off the names of Snow White's seven woodland companions, by revealing the make and model of the car in which the Austrian archduke Francis Ferdinand rode to his assassination, and by recalling the eye on which the Hathaway man wears a patch. I wanted to become the life of the party, so in front of the others I offered to bet him $5 that if you drop a $5 bill from five feet, it will land with Lincoln face up. How much of a fool was I?

Hun Dread

I once met a Mongolian who was trying to better himself by teaching himself English. He was surprised to find that the initial letter of the English alphabet is rarely seen in certain contexts. For example, when you are counting from zero, what's the first number you come to that has an *a* in it?

U.S. History

In graduate school I was taught to take nothing for granted. I came to scrutinize even the most widely held beliefs in a fresh way. Eventually I was able to convince myself that the Pope is Catholic and that the color of George Washington's white horse is white. But who did I find buried in Grant's tomb?

Write Pretty, Eh?

What is QWERTYUIOP? Why is QWERTYUIOP what it is? What apt ten-letter word can you make from QWERTYUIOP? You may repeat letters, so you need not use all of them.

ANSWERS

The Well-Hung Natives

Remember the primitive Island of Noose, where if a woman cheats on her husband she tells everyone except him, and if the husband finds out, he hangs her at midnight in the public square? Well, the missionary's speech to the natives about cheating on the island results in the death of all the adulteresses. It turns out that if *n* women cheated, they are all hanged on the *nth* night. Let me explain the case of $n=1$ and the case of $n=2$.

Consider what happens if only one heathen has been unfaithful. Before the missionary gives his speech, everyone except the husband knows that she cheated on him. Because the sad sap doesn't know of any cases of adultery, the missionary's speech points the finger at his wife.

Now imagine what happens if two women have been unfaithful. Everyone except the two wronged men knows of two cheaters. And each of the cheated men knows of one cheater: the other one's wife. This means that on the night of the missionary's speech, all the natives sleep soundly because his parting words do not immediately communicate anything the primitives didn't already know. The next day, however, the two cheated men are surprised to see that the other one's wife is not hanging in the public square. Each man realizes that if the other man's wife were the only cheater, she would have been hanged that night. Consequently, there must be more than one promiscuous squaw. But each man would know the name of another adulteress if it were anyone but his own wife. Hence each man knows that his own spouse has cheated on him. So both women were hanged at 12:00 P.M. on the second night.

What happens if the missionary was lying and there has actually been no cheating on the island? In that case, each husband will think that his wife is the culprit. At midnight every husband and his spouse will show up in the public square. Each husband will be shocked to see the other couples. No one will be hanged because they can all assure one another that their wives have not been engaging in extramarital congress.

Crocked Crypton

If you try to solve the problem by working backward, you may get hopelessly bottled up. There was a point near the end of the drinking spree when I polished off the eleventh beer, leaving only the wine and the twelfth beer. Where had the eleventh beer been? Suppose I had proceeded clockwise around the circle. Was the eleventh beer between the wine and the twelfth beer or between the twelfth beer and the wine? You need to test each situation in order to determine whether a clockwise count of thirteen from the eleventh beer gives the twelfth beer. This approach becomes needlessly unwieldy for earlier points in the drinking spree when more bottles remained.

The easiest way to solve the problem is to work forward. Write the numbers 1 through 13 in a circle. Assume I begin with 1; put an X through it to indicate I drank it. Now count clockwise another thirteen numbers. That takes me to 2. Put an X through it. Count clockwise another thirteen numbers and cross out the result. Keep repeating this procedure until you are down to one number, which represents the bottle of wine. the wine turns out to be 9. In other words, to drink the wine last I started with the beer that is five places clockwise from the wine. (I want you to know that I was sexcessful in winning the foxy woman's heart.)

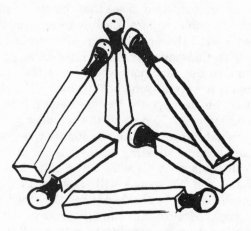

Crypton's Answer

Matching Wits

Four equilateral triangles can be made from six unbroken matches by arranging them in three dimensions. Use three matches to form a triangle in the plane of this paper. Then put one match at each angle of the triangle so that they come together at a point above the paper.

Correspondents had a field day with "Matching Wits." Several readers, including Richard Foy of Redondo Beach, California,

Readers' Solutions

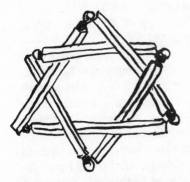

Richard Foy, Rick Friedberg

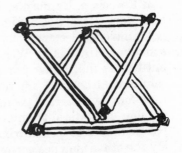

Rita D. Widdison

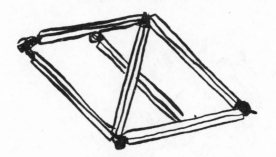

Mark Mesick

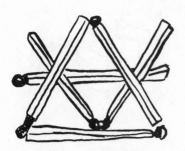

Warren E. Rupp

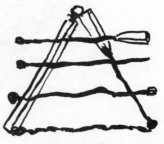

Glen Tarachow

and Rick Friedberg of Susanville, California, suggested that the six matches be arranged in the form of a Star of David. Indeed, this arrangement is more energy efficient than my solution because it yields eight equilateral triangles, two big ones and the six points of the star.

Alternative solutions were also submitted by Rita Widdison of San Francisco, California, Warren E. Rupp of Mansfield, Ohio, and Mark Mesick of St. Paul, Minnesota.

My favorite letter about the triangles came from Glen Tarachow of Milwaukee, Wisconsin. "Please excuse my boldness, but I think that I have proved that the puzzle has an alternative solution. Although I am young (16) and have not studied more than four semesters of geometry, I feel that my method of creating four equilateral triangles from six matches is without fault." His method is to form a triangle from three matches and then to lay the other matches across the triangle parallel to its base. As long as the last three matches are parallel to the base it doesn't matter where they are. The virtue of this approach is that it yields an infinite number of solutions.

The Lighter Side

The inspired way to make a Christian cross with five matches is to jam four of them in his mouth and light them with the fifth one. (The joke as I heard it involved a "Nazi cross," but that phrase flags the punchline because "swastika" is what you'd normally say.)

No Great Shakes

The SHARP JOKER is the anagrammatic nickname of puzzle fiend Josh Parker. To solve a problem such as this one (or "The Well-Hung Natives" problem), in which you don't seem to be given enough information, it is best to start with an extreme case. Consider the person who shook eight hands. This glad-hander pumped the hand of everyone except, of course, himself and his mate. But we know that someone at the party didn't shake hands at all. Who was it? Everyone except the glad-hander's mate participated in at least one handshake, so that by process of elimination we know that the glad-hander's mate is the one who didn't shake any hands.

Now take the person who pumped seven hands (the hands of everyone except himself, his mate, and the person who shook zero hands). The only person who can shake only one hand is the mate of the person who shook seven hands, because everyone else is committed to shaking no hands or more than one hand. By the same token, the person who shook six hands is coupled with the person who shook two hands, the person who shook five hands is coupled with the person who shook three hands, and the person who shook four hands has a mate who also shook four hands. The Sharp Joker is reported to have shaken the hands of the same number of people as someone else did. The only number that repeats is four, and so he shook four hands. Therefore the hostess shook four hands too.

A Huemuris Question

Generations of puzzle freaks have thought that the bear has to be at the North Pole, with the man standing 100 yards to the south. Only then could he walk 100 yards due east and be due south of the beast. This means that the bear is a polar bear; hence, it is white.

The preceding solution was accepted for years until Raymond

Smullyan argued ingeniously in the pages of *Scientific American* that there is an infinite number of solutions. Smullyan repeated his argument in a book whose title I often forget, *What Is the Name of This Book?* "It could be, for example, that the man is very close to the South Pole on a spot where the polar circle passing through that spot has a circumference of exactly 100 yards, and the bear is standing 100 yards north of him. Thus if the man walks east 100 yards, he would walk right around that circle and be right back at the point he started from. So that is a second solution."

A third solution that Smullyan suggests is one in which the man stands at a point where the polar circle has a circumference of 50 yards, so that his stroll would take him twice around the circle back to where he started. And in yet a fourth possible answer, the polar circle has a circumference of 100 yards divided by 3. There the hunter walks three times around to return to his starting point. You get the idea. The teaser is solved by an infinite number of points through which the circumference of the polar circle is equal to 100 yards divided by any positive integer.

As clever as Smullyan's mathematical analysis is, it falls short of solving the problem because it is based on faulty mammalogy. Ther are no polar bears anywhere near the South Pole. In fact, aside from zoos, the nine-foot beasts are seldom found south of the latitude that runs through Liverpool, England.

Remarkably, the whole problem of the hunted bear is flawed because polar bears, in spite of their name, do not live at either pole. The North Pole is too frozen for life to thrive on which the bears could feed. It is unfortunate that the problem collapses because of its very charm: the color of the bear.

One of my readers, Sam Bowman of Eagle, Alaska, objected to the shotgun in my formulation of the puzzle. "Polar bears are not usually hunted with a shotgun," he wrote, "and your target would be quite unharmed at 100 yards. In this part of the country a 12-gauge shotgun loaded with 00-buck is considered to be the ultimate bear insurance, but this is strictly for close-range defense. Otherwise, your column is very interesting."

Summary Judgment

Churchill, Hitler, Mussolini, Roosevelt, and Stalin were in effect participating in a variation of an old calendrical magic trick, which is best done at parties. The magician asks for a volunteer, to whom he says, "You must be in your thirties" (or whatever the person's age appears to be). He continues: "Don't tell me your exact age. Just write down on a piece of paper the year you were born." The magician asks for a second volunteer, who is told to think of a great event in his personal life. He is instructed to write the year of the great event on the slip of paper with the other number. The first volunteer is then asked to write down how old he'll be at the end of the current calendar year. Finally, the second volunteer writes down what anniversary of the great event this year will mark. The conjurer asks the second volunteer to compute the sum of the four numbers. Before the computation is completed, the magician announces the sum. The magician is right; how did he do it?

	CHURCHILL	HITLER	ROOSEVELT	MUSSOLINI	STALIN
YEAR OF BIRTH	1874	1889	1882	1883	1879
AGE IN 1944	70	55	62	61	65
YEAR TOOK OFFICE	1940	1933	1933	1922	1924
YEARS IN OFFICE	4	11	11	22	20
SUM	3,888	3,888	3,888	3,888	3,888

The calendar trick for parties works on the simple arithmetical fact that the year in which a person was born plus his age at the end of the current year is equal to the current year. Similarly, the year of an event plus the anniversary that the current year marks of that event equals the current year. Before the magician does the trick, he doubles the current year.

A bizarre letter from Del O. Scheidler of Eagan, Minnesota, asked me why I didn't complete the chart that accompanied "Summary Judgment." He finished it for me by adding Tojo's vital statistics and by changing Mussolini to Il Duce. Scheidler then makes the remarkable observation, "To find the supreme ruler (?), take the first letter of each name in the chart." Christ is a popular figure in many letters I receive. One man wrote me that Christ was an acronym for "Can He Rise in Society Today?"

Memory Lane

Hubert and Samantha's interchange is an enviable example of perfect communication. Hubert has provided Samantha with mnemonic devices for keeping the trigonometric functions straight and for remembering the expansion of π. The sine of an angle in a right triangle is the length of the opposite side divided by the length of the hypotenuse. The cosine is the length of the adjacent side divided by the length of the hypotenuse. The tangent is the length of the opposite side divided by the length of the adjacent side. The three functions can be summed up by the sequence of letters SOHCAHTOA, an acronym for "Sine Opposite Hypotenuse, Cosine Adjacent Hypotenuse, Tangent Opposite Adjacent." The school marm who taught me trig told me I could remember SOHCAHTOA if I pronounced the acronym as if it were the name of an Indian chief. But how does one avoid confusing big chief SOHCAHTOA with big chief SAHCOHTOA?

Hubert has a cleverer mnemonic. Imagine SOHCAHTOA is an acronym for the very memorable phrase: "Sex on Holidays Can Advance Happiness to Outrageous Amplitudes."

What about π? The decimal expansion of π to four places is 3.1416. The number of letters in each successive word of Hubert's remark, "Yes, I know a number," gives the digits of π in order.

Suppose Samantha had to know π to 30 decimal places. Hubert would have told her to keep in mind a French verse, which is quoted in Helen A. Merrill's *Mathematical Excursions* (Dover, 1957):

Que j'aime à faire apprendre un numbre utile aux sages!
Immortel Archimede, sublime ingénieur,
Qui de ton jugement peut sonder la valeur?
Pour moi ton problème eut de pareils avantages.

Again, it is the number of letters in each word that's important. The verse can be translated:

How I want to teach a useful number to wise men!
Immortal Archimedes, sublime engineer,
Who can sound the value of your judgment?
For me your problem had parallel advantages.

Merrill also offers an English poem that gives π to 12 decimal places:

See, I have a rhyme assisting
My feeble brain, its tasks ofttimes resisting.

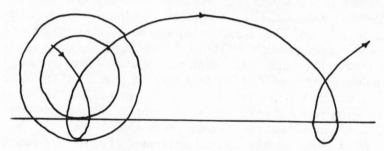

Trajectory of Point on the Flange

Backwald Tlain

No, he was not deranged. Consider first the case of a train whose wheels rest directly on the track. The point of contact between a wheel and the track is always instantaneously at rest, regardless of the speed of the train. If the point were not at rest, the wheel would slide along the track rather than roll along it. The other points of the wheel are momentarily rotating around the instantaneous rest point.

The idea of instantaneous rest should not freak you out. Suppose you throw a softball into the air. For the ball to return to the earth, as it surely will, its velocity away from you must change from a positive number to a negative number. Hence, at some point (namely, the top of the ball's trajectory), the velocity is zero and the ball is momentarily at rest.

The wheels of most trains have an extension called a flange that keeps the wheels on the track. The section of the flange that extends below the point of contact of the main body of the wheel and the track also rotates around the contact point. As a result, this part of the flange is moving backward. If the flange is one-tenth the width of the main body of the wheel and the train moves forward with a velocity of forty miles per hour, the bottom edge of the flange moves backward with a velocity of four miles per hour.

Lake Chargoggagoggmanchauggagoggchau-bunagungamaugg

The second fish is the odd man out because it is the only one without a twin. This problem is based on an actual IQ-test question, which is reprinted in Michael Steuben's puzzle column in *Capital M*.

How to Spot a Good Secretary

There are six possible color combinations to consider: (1) five white dots, (2) four white dots and one black dot, (3) three white dots and two black dots, (4) two white dots and three black dots, (5) one white dot and four black dots, and (6) five black dots. The fact that all five applicants stood up rules out the first four possibilities; in each case there is at least one person who would not have seen three black dots and hence would not have stood up.

The fact that someone raised his hand eliminates the sixth possibility. If all the dots were black, each contestant would be looking out at four black dots. Call one of the contestants Gail. She would realize that each contestant would see three black dots regardless of whether her own dot was black or white. Hence she cannot deduce the color of her own dot. The same reasoning applies to the other contestants; none can figure out the color of his own dot.

That leaves only the case of four black dots and one white dot. The person whose dot is white would not be able to determine the color for the same reason Gail couldn't. Now consider Paul and Matt, two of the four applicants whose dots are black. Paul is looking at one white dot and three black dots. Paul realizes that Matt had to see three black dots in order to stand. But apart from Paul's own dot, Matt can only see two black dots. Hence Paul knows his own dot is black, too.

Any of the four players who had black dots could have made this deduction. Paul was the fastest, and it took him a few minutes because he had to work through the reasoning that I've just presented. The game is not fair because the player whose dot is white does not have enough information to determine the color.

Three correspondents, Virgil May of Corona, California, Gerald Silver of New York, New York, and Easy Jones of Kirksville, Montana, offered a better solution to "How to Spot a Good Secretary." "I agree," Silver wrote, "that there are only two possibilities: (a) four black dots and one white dot and (b) five black dots. I assume that each of the five players is smart enough to realize this. Therefore, if any of the players saw a white dot, that person would immediately raise his hand and announce that he had a black dot. However, nobody raised his hand for a few minutes. I reason that this was because no one saw a white dot. One person realized more quickly that everyone had a black dot and raised his hand." In that case, May concluded, "the game is fair to everyone."

Deplaning with the Motion-Discomfort Container

In describing the return of a rocket, a spokesman for NASA said, "When the craft *deboosts* into the earth's atmosphere . . ." And on the Long Island Rail Road the conductors are fond of announcing, "The first two cars of the train will not *platform* at Mastic-Shirley."

The most barbaric example I can cite is the rent-a-car clerk who told me that the car should be *degaraged* by the rear exit. The absurd *degaraged* at least has a logological claim to fame: It is a palindrome. Indeed, with nine letters, *degaraged* is right up there with the longest legitimate English palindromes, *evitative* and *redivider*.

Sesquipedalian Fun

The first paragraph of the puzzle can be translated with the aid of Dmitri A. Borgmann's *Language on Vacation* (Charles Scribner's Sons, 1965) and *Mrs. Byrne's Dictionary* (University Books, 1974), a splendid lexicon of extraordinary words. The paragraph means: "I want to rejoin the ranks of those who love nibbling on women's earlobes, so on the advice of my rigid doctor I took time off from my philosophical and psychological work on the twenty-six-sided solid in order to recover from the depression from my long hospital stay for an operation in which my gall bladder was connected to my hepatic duct and an operation in which two loops of my jejunum were linked." Lora and I drove to Massachusetts and rented a cabin on Lake Webster. The fourth-from-last fish I caught was an inedible tropical fish from Hawaii. (Don't ask me what it was doing in New England.) So we ate the leftovers from the meals of the leftovers from the meals of the past two weeks. On the drive home I proposed we see which one of us could think of the longest word. At first Lora dismissed my game about long words as bunk, but after I chided her for categorizing the game as trivial, she agreed with honorableness to play.

I responded to Lora's forty-five-letter word with the 1,185-letter name for the protein that makes up a strain of the tobacco mosaic virus:

ACETYLSERYLTYROSYLSERYLISOLEUCYL
THREONYLSERYLPROLYLSERYLGLUTAMI
NYLPHENYLALANYLVALYLPHENYLALAN
YLLEUCYLSERYLSERYLVALYLTRYPTROPH
YLALANYLASPARTYLPROLYLISOLEUCYL
GLUTAMYLLEUCYLLEUCYLASPARAGINYL
VALYLCYSTEINYLTHREONYLSERYLSERY
LLEUCYLGLYCYLASPARAGINYLGLUTAMI
NYLPHENYLALANYLGLUTAMINYLTHREO
NYLGLUTAMINYLGLUTAMINYLALANYLA
RGINYLTHREONYLTHREONYLGLUTAMINY
LVALYLGLUTAMINYLGLUTAMINYLPHENY
LALANYLSERYLGLUTAMINYLVALYLTRYP
TOPHYLLYSYLPROLYLPHENYLALANYLPR
OLYLGLUTAMINYLSERYLTHRONYLVALYL
ARGINYLPHENYLALANYLPROLYLGLYCYL
ASPARTYLVALYLTYROSYLLYSYLVALYLT
YROSYLARGINYLTYROSYLASPARAGINYL
ALANYLVALYLLEUCYLASPARTYLPROLYL
LEUCYLISOLEUCYLTHREONYLALANYLLE
UCYLLEUCYLGLYCYLTHREONYLPHENYL
ALANYLASPARTYLTHREONYLARGINYLAS
PARAGINYLARGINYLISOLEUCYLISOLEUC
YLGLUTAMYLVALYLGLUTAMYLASPARAGI
NYLGLUTAMINYLGLUTAMINYLSERYLPRO
LYLTHREONYLTHREONYLALANYLGLUTA
MYLTHREONYLLEUCYLASPARTYLALANY
LTHREONYLARGINYLARGINYLVALYLASP
ARTYLASPARTYLALANYLTHREONYLVALY
LALANYLISOLEUCYLARGINYSERYLALA
NYLASPARAGINYLISOLEUCYLASPARAGIN
YLLEUCYLVALYLASPARAGINYLGLUTAMY
LLEUCYLVALYLARGINYLGLYCYLTHREON
YLGLYCYLLEUCYLTYROSYLASPARAGINY
LGLUTAMINYLASPARAGINYLTHREONYLP
HENYLALANYLGLUTAMYLSERYLMETHIO
NYLSERYLGLYCYLLEUCYLVALYLTRYPTO
PHYLTHREONYLSERYLALANYLPROLYLAL
ANYLSERINE.

I first came across the word for the protein in Borgmann's splendid book *Beyond Language* (Charles Scribner's Sons, 1967). The word is not a product of Borgmann's diabolical logologistic mind but a technical term listed under the molecule $C_{785}H_{1220}N_{212}O_{248}S_2$ in the formula index of *Chemical Abstracts* (volumes 56–65). The protein consists of 158 amino acids, and the name is constructed from the names of the amino acids in the order in which they appear in the molecule.

In 1967 Borgmann predicted that in a year or two *Chemical Abstracts* would include a 1,385-letter name for human growth hormone $C_{973}H_{1143}N_{247}O_{306}S_7$. Nevertheless, it seems that *Chemical Abstracts* changed its nomenclature so that a chemical name is not spelled out laboriously but abbreviated according to a special terminology. In any event, Borgmann's prediction is undoubtedly out of date, because the formula index of *Chemical Abstracts* for 1972–1976 includes seventy-five molecules with more carbon atoms than the mere 973 such atoms in human growth hormone. The greatest number can be found in the human protein albumin, which has the formidable formula $C_{2936}H_{4625}N_{787}O_{888}S_{41}$. I have not been able to find in print, however, the names of any of these giant molecules. Nevertheless, there are more than one thousand volumes of *Chemical Abstracts* in the library (imagine the staggering number of volumes of periodicals from which *Chemical Abstracts* abstracts!), so that a megaletter word might be in there somewhere. Have a look!

The Book of Lists (William Morrow and Company, 1977) cheats when it claims, without spelling the word or citing a source in which it is spelled, that the longest word in the English language is a bovine protein that has 3,600 letters. I defy anyone to send to me a photocopy of a page from a book or a periodical that includes such a word. I'd also like to learn of fictitious words that are monstrous. (You'll have to do better than one hundred letters, of which there are several examples in James Joyce's *Finnegans Wake*.) Please send me a photocopy of any you find.

The Guinness Book of World Records (1975) spells out all 1,913 letters of the word for tryptophan synthetase A protein, $C_{1289}H_{2051}N_{343}O_{375}S_8$. I am somewhat troubled by Guinness's claim that the protein has the longest name of any chemical. I have not seen the word spelled out anywhere except in *Guinness*. The word is certainly not in *Chemical Abstracts*. If it has never appeared in a chemistry book or a journal, it seems to me that it isn't kosher to hire a chemist to spell out the word from the formula just so the word can be quoted in *Guinness*.

In any event, if you're going to take the trouble of spelling a chemical name that has never been written out before, you might as well go after the biggest game. That is what *Word Ways: The Journal of Recreational Linguistics* did. Jeff Grant of Hastings, New Zealand, provided the journal's editor, A. Ross Eckler, with the 3,641-letter name of the protein bovine glutamate dehydrogenase. Two parts of the word, "glutamylasparaginyl" and "glutaminyl," may not be accurate because of uncertainties in the chemical structure. One proposed resolution of the uncertainties would unfortunately lead to a name with only 3,639 letters. What is amazing about the article in *Word Ways* is that after it spells the word, it gives a phonetic pronunciation, which includes the amusing sequences "Jew oy" and "joy pain knee cheek" (what Eckler calls the hypochondriac's lament). Eckler maintains that the word, which contains 577 sounds in 253 syllables, can be pronounced in ninety seconds.

Denis Farrant of Hollywood, California, wrote me that the longest English word is *smiles*; between the two *s*'s there's a mile.

A Grave Pattern

Although assassination may be the leading cause of death of presidents in office, natural causes have brought down as many presidents as bullets have. Abraham Lincoln, James A. Garfield, William McKinley, and John F. Kennedy were assassinated, whereas William Henry Harrison, Zachary Taylor, Warren G. Harding, and Franklin D. Roosevelt died naturally.

Taylor is often overlooked because his death does not fit into the familiar numerological pattern governing the fate of every president elected at twenty-year intervals since 1840. Each president died in office. Moreover, and what I have not seen pointed out before, each one died in an odd-numbered year, a tragedy that President Reagan narrowly avoided. Harrison, elected in 1840, died in 1841, reportedly of pleurisy and pneumonia. Lincoln, elected in 1860, was assassinated in 1865. Garfield, elected

in 1880, was assassinated in 1881. Harding, elected in 1920, died in 1923, reportedly of apoplexy (rupture of a brain artery), pneumonia, enlargement of the heart, and high blood pressure. Roosevelt, elected in 1940, died in 1945 of a cerebral hemorrhage. And Kennedy, elected in 1960, was assassinated in 1963. So of the presidents elected at these twenty-year intervals, the leading cause of death is assassination. The only other president to die in office is Taylor, elected in 1848, who died in the White House in 1850, reportedly of bilious fever, typhoid fever, and cholera morbus.

Start

End of First Interval

End of Second Interval

Riding Between the Lines

Since the army and the courier both move at constant speeds, the ratio of the distance the army moves in any interval of time to the distance the courier moves in that same interval is a constant. Consider two intervals of time. The first interval is from when the courier sets out to when he reaches the front of the army. The second interval is from when he reaches the front of the army to when he reaches the rear of the army.

Now draw a picture of the positions of the army and the courier at the end of each interval of time. Let X be the distance in miles that the army moved in the first interval. Then $5 + X$ is the distance the courier moved in that same interval. So the ratio is $X \div (5 + X)$. In the second interval, the army moved a distance $5 - X$ (because it must move a total of 5 miles for the rear to get to where the front was), and the courier traversed a distance X. So the ratio is $(5 - X) \div X$. Setting the two ratios equal and a little algebra yields a value for X of $5 \div \sqrt{2}$, or about 3.54. Therefore, in the two intervals the courier covered a distance of $5 + X + X$, or 12.08 miles.

Dead Precedents

You would not say, "Three out of the first five presidents" did anything unless the fifth was one of the three. Otherwise, you would say "Three out of the first four." Mrs. Smith's son did not need to know the name of the fifth president because in front of him was a list of the presidents in order.

Underground Expression

Why, 125. The number sequence expresses the uptown express stops on the Lexington Avenue subway line in New York City. I apologize to those readers who have never set foot in the Big Apple.

The next number in the math jock sequence is 12. The sequence represents the number 8 expressed in successive bases, starting with base 2. In base 6 the number 8 is 12.

The next card is the five of clubs. The denominations of the cards are in reverse alphabetic order. The suits alternate between clubs and diamonds.

The subway-stop sequence has not made me popular. Many readers put in a considerable amount of time trying to find a "logical" answer. Correspondents were able to justify 102, 112, and 126.

"After studying your problem, I determined there was a numerical sequence," wrote Randy Emery of Missoula, Montana, "only to find that I was supposed to be on a wild goose chase, having never been to New York City. However, if you subtract 14 from 59 and then 42 from 86, you will presumably get 45 and 44 respectively. The logical step, naturally, would be to assume that the next difference is 43." Therefore, the next number is 43 plus 59, or 102.

"Foul!" wrote Don R. Dalton of Bloomington, Indiana. "So that my efforts shall not have been in vain, I submit 122 as a solution for your amusement." Dalton arrived at 122 by first taking the difference of successive numbers in the sequence and then taking the difference of the differences:

$$\underline{14} \qquad \underline{42} \qquad \underline{59} \qquad \underline{86} \qquad \underline{122}$$

$$42 - 14 = \underline{28} \quad 59 - 42 = \underline{17} \quad 86 - 59 = \underline{27} \quad 122 - 86 = \underline{36}$$

$$28 - 17 = \underline{11} \quad 27 - 17 = \underline{10} \quad 36 - 27 = \underline{9}$$

Dennis Barnett of Hollywood, California, tried to justify 126 by taking the difference between 14 and each number in the sequence:

$$42 - 14 = 28 = 4 \times 7$$
$$59 - 14 = 45 = 5 \times (7 + 2)$$
$$86 - 14 = 72 = 6 \times (7 + 2 + 3)$$
$$126 - 14 = 112 = 7 \times (7 + 2 + 3 + 4)$$

The connection between the first two equations, however, is somewhat obscure.

A Relatively Niece Photo to Make You Cry Uncle (Aunt You Glad I Didn't Work in Cousin Too?)

The best way to solve the problem is to work backward from the cousin's mother. It is easiest to look at the answer from the point of view of the friend. My cousin's mother is my aunt. The brother of my aunt is my father, because my comment on the second photo revealed that my only aunt has only one brother. The only grandson of my father is my son. The uncle of my son is my wife's brother, because my comment on the first photograph revealed that I have no brothers. The only cousin of my wife's brother is my wife's cousin. The father of my wife's cousin is my wife's uncle. The only niece of my wife's uncle is, at last, my wife.

Biff and Buff

Newton's law of action and reaction states that for any action there is an equal and opposite reaction. When Buff grasps the rope above him and pulls down on it with a given force, the rope responds by pulling up on him with the same force. Since the rope does not stretch, it acts like a rigid rod; the tension at Buff's end (produced by his weight and his pulling on it) is instantaneously transmitted over the pulley to Biff's end. There the tension pulls up on Biff with the same force it exerts on Buff, and so Biff moves closer to the pulley at the rate at which Buff climbs toward it. Sometimes it pays to be lazy. You can have your cake and eat it, too.

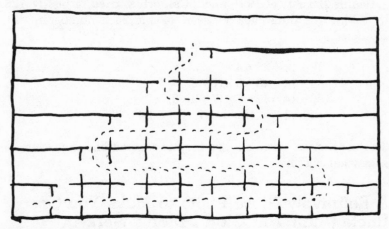

Feat of Clay

I visited thirty-one rooms in Marc Swenice's gallery, which is filled with NEW CERAMICS (an anagram of his name). Once you enter a room in one of the horizontal rows (other than the first row consisting of one room), you will have to miss one room in that row in order to be able to move on to another row. Hence you will not visit five rooms, one in each row other than the first row.

Lincoln's Downfall

I was not a fool at all. The bet is a sure one because Lincoln appears on both sides of a $5 bill. He of course looks out from the front of the bill. And on the back a little statue of him is inside the Lincoln Memorial. Look closely, now. The statue is only three-sixteenths of an inch high.

Even if Lincoln did not appear on the back of the bill, the bet would not be a foolish one because I'd have a 50 percent chance of winning. Of course, offering an even bet would not make me the life of the party.

For your edification, the Seven Dwarfs are named Bashful, Doc, Dopey, Grumpy, Happy, Sleepy, and Sneezy. The Archduke took his last ride in a 1912 *Graf und Stift*. And the Hathaway man has a patch on his right eye.

Gerald Johnstone of Santa Ana, California, thanked me for the bet about how Lincoln can be found on both sides of a $5 bill. "At the rate it's working," he wrote "I should be able to retire in three years."

Craig Brownwell, of Cook Station, Montana, took issue with my assertion that even if Lincoln did not appear on the back of the bill, the bet would not be a foolish one because there would be a 50 percent chance of winning. "You would actually have more than a 50 percent chance of winning if Lincoln only appeared on the front side of a $5 bill," Brownwell wrote. "People usually put a bill in their wallet face forward. This means that the crease would tend to make the bill land face up when dropped from a sufficient height. You should try this yourself. This would not, of course, be the case with a crisp, uncreased $5 bill from the bank. But the chances are that he would pull the bill from his wallet (naturally you would let him use his own bill)."

R. S. Craggs of West Hill, Ontario, warned me that he would win the bet about Lincoln if he could choose the bill. "You did say a '$5 bill,' not an 'American $5 bill.' The $5 bill that I currently have in my wallet bears a picture of Sir Wilfred Laurier on the face side and a picture of some kind of fishing boat on the reverse side. I doubt if I would find a replica of Abraham Lincoln concealed anywhere in the engraving even with the aid of the most powerful microscope." All I can say is "Fifty-four Forty or Fight."

Hun Dread

The first number with an *a* in it is one thousand. If you thought that the answer is "one hundred and one," you weren't paying sufficient attention to the old schoolmarm who taught you how to count. The number is not "one hundred and one" but "one hundred one." Remarkably, although *a* is the third most common letter in the English language (after *e* and *t*), an *a* does not appear in the first 999 numbers (actually the first 1,000 numbers if you include zero).

Gerald Johnstone commented on the Mongolian I met. "If I remember correctly, he's the one who presumably spent the better part of an hour counting from zero in search of a number with the letter *a*. If he had only changed his direction of travel, he would have found more time for his English studies with the discovery of negative one. I hope this doesn't fall in the category of *the hair in the egg*." Mike Miller, of an undisclosed location, made the same point.

Palmer Cone of Elkhart, Indiana, wrote that the Mongolian would be even more perplexed if he counted until he came across a *b* or a *c*. "Am I correct in stating that you must count to a billion in order to find a *b* and to an octillion in order to encounter a *c*?"

U.S. History

It turns out that Grant's Tomb is occupied not only by Ulysses but also by his wife Julia—in separate but equally big and noble coffins. The vault should be called Grants' Tomb.

Now, would you believe me if I told you that the Canary Islands were not named after the songbird? Of course you would. And you'd be right. The islands were named for the large dogs that romped there. The Latin name for the Canary Islands is *Canariae insulae*, "Isles of Dogs." Next I'll be telling you that England, the country the United States broke from, also has a Fourth of July. And you thought their calendar went directly from July 3 to July 5?

Before I'm through with you, I'll make you a total skeptic. If you can't trust polar bears, the Canary Islands, and Grant's Tomb, what can you take at face value? I hope you'll send me things that people take for granted but that are just not so.

Two correspondents challenged my diction in "U.S. History." John Sanfilippo of Santa Clara, California, wrote: "You and Groucho Marx made the same mistake. If you saw the old *You Bet Your Life* game show, you'll recall that when Marx had two guests who seemed to be stupid, he'd always ask the same question, 'Who is buried in Grant's tomb?' The guests wouldn't know the answer, so Groucho would tell them that Grant was buried in Grant's tomb. But if you look in a dictionary, you'll see that bury is defined as to put into the earth. We both know that Ulysses and Mrs. Grant were not directly put into the earth but entombed. The proper question should be: Who is entombed in Grant's tomb?"

Mark Lynner of Watford City, North Dakota, made the same point. "I love your column. But did you have to goof up so soon in your career? The truth is nobody is buried in Grant's tomb. People are entombed in tombs and buried in graves." Oh, well.

Write Pretty, Eh?

QWERTYUIOP is the top row of alphabetic keys on an ordinary typewriter. Why are the keys ordered this way? In early typewriters the keys often jammed. As a result the keys were apparently arranged so that the ones that follow each other on the keyboard correspond to letters that are unlikely to follow each other in actual usage. (This does not explain, however, why the key for *r* follows the key for *e* because *er* is a common word ending.) The contrived order has remained in spite of the fact that on most modern typewriters adjacent keys no longer jam.

Seymour Papert, an expert in artificial intelligence at MIT, has coined the term "QWERTY phenomenon" for any practice to which people are committed that has no rational basis beyond its historical roots.

The ten-letter word constructed from QWERTYUIOP? TYPE-WRITER.

By the way, "WRITE PRETTY, EH?" is an anagram of THE TYPEWRITER.

MORE MAIL

Howard B. Julien of Eureka, California, challenged several of my answers: "I would bet you that the moon was made of green cheese. The fool would be you for accepting a bet that could be easily reneged.

"The bear puzzle has been one of my favorites for years, but the appearance of a shotgun (as opposed to a high-powered rifle) changes the reasoning. Even if polar bears lived at the North Pole, the event described could not have taken place. A shotgun could only kill a bear at very close range. All possible North Pole solutions would involve either shooting from too great a distance to kill or circling the bear from so close that he wasn't being given a chance to get away.

"The event must have taken place at the South Pole. Although there are no bears native to Antarctica, nothing in your description of the event prohibits the importation of a bear specifically for the event. The members of the scientific community who do research at the Pole would prohibit the killing of any species of bear whose population in the wild is declining (polar, Kodiak, grizzly, honey, and so on). That leaves the brown bear as the only possibility.

"This of course leads to another paradoxical question: What color is a brown bear? Black, the correct answer to your puzzle."

Bobby Rodriguez of Green Bay, Wisconsin, offered a definition of recombinant DNA: "designer genes."

CHAPTER 9

The Logistics of Armchair Ornithology

IF AFTER A SUCCESSION of calamitous romances you concluded that all intimate relationships are destined to fail, you would be doing what the scientist does routinely: generalizing from a handful of experiences to all future cases. Called induction, this kind of extrapolation is made in all branches of science, from ornithology and ichthyology to volcanology and particle physics. When a bird watcher crouched in a duck blind spots a large flock of white swans and concludes that all swans are white, he is making an inductive inference. And so is the physicist who, after scrutinizing a finite number of cases, concludes that all objects gravitationally attract each other with a force directly proportional to the product of their masses and inversely proportional to the square of the distance between them.

The outstanding feature of an inductive inference is that it could turn out to be false. The next romance could go smoothly, the next swan could be black, and the next pair of objects could fail to attract each other gravitationally. (I'm following the philosophical literature in assuming that all swans are white; actually, some swans *are* black.) There is nothing parodoxical, however, about this feature of induction. The future cannot be logically inferred from the past. What has been observed imposes no logical constraints on what will be observed.

Nor does the possible falseness of an inductive inference impede the progress of science. If an ornithologist happens to observe a red bird that is like a swan in every respect except color, he can either withdraw his claim about all swans being white or he can decide that the red anomaly is not a swan but of some other family.

115

All swans are white.

The real problem with induction comes from another source: Observed regularities are sometimes accidental rather than causal. Suppose you asked three people in a large restaurant if they were married, and they replied in the affirmative. You would not conclude that all the people in the restaurant were married.

Or consider the mind-expanding material you're now reading. Provided you've been reading this book from the beginning, every word you've read you've read prior to completing this sentence. Yet you couldn't generalize from this and expect to read every word of mine you'll ever read before finishing this sentence (or if you do, you'd better put the book aside immediately, because you have only nine words left before being exposed as a fool). Yet if you hooked up a few pennies to a battery and found that they conducted electricity, common cents would tell you that all pennies conduct electricity.

The problem is not in recognizing which observed regularities are accidental and which are causal. No one would doubt that some regularities (the marital status of the patrons of a restaurant, for example) are accidental and that some (the conducting pennies) are not. The difficulty is in formalizing the recognition. Philosophers have spent centuries trying to spell out in unambiguous language what it is that makes some regularities causal and others accidental.

Do you see a way out? Perhaps you would argue that there is an underlying uniformity to nature that makes certain regularities support inductive generalizations. Such things as the conductivity of pennies, the color of birds, and the gravitational attraction of objects are manifestations of this fundamental uniformity, whereas the marital status of diners and the order in which you read are not.

Invoking an underlying uniformity of nature serves not to solve the paradox of induction but to obscure it. Although cast in fresh language, the same paradox continues to perplex us: What is it about the regularity of nature that manifests itself in the regularity of gravitationally attracting objects but not in the regularity of marital status? A popular but ill-conceived way of approaching the paradox is to maintain that accidental regularities involve spatial or temporal references, whereas natural regularities, supposedly universal and eternal, make neither kind of reference. In this way, the regularities cited about marriage and word order can be dismissed as accidental because of the reference to a particular place (a certain restaurant) and to a temporal notion *(when* you read what).

Nevertheless, this explanation ultimately collapses because it applies neither to accidental regularities nor to natural ones. Let me adjust the example of the words you've read. Each word you've read in this chapter starts with a letter other than z, yet you would not therefore expect every word of mine you'll ever read to start with a letter other than z. This observed regularity is clearly an accidental one, but it does not involve a spatial or temporal reference. Moreover, the statement "All swans in

North America or elsewhere are white" obviously expresses a nonaccidental regularity in spite of its conspicuous spatial reference.

Another attempt to explain the problem of induction takes its cue from the fact that induction has worked in certain situations in the past and thus should be expected to work in similar situations in the future. Suppose ornithologists in the eighteenth century had concluded after observing a flock of black ravens and pink flamingos that "All ravens are black" and "All flamingos are pink." (I *know* some verge on scarlet; indulge me.) With the passage of time, these two predictions have not been discredited; in fact, they have gained credibility because many more ravens and flamingos have been observed.

Doesn't the success of these predictions point to an underlying uniformity of nature with regard to the color of birds? To put it another way, shouldn't the experience with ravens and flamingos increase our confidence in concluding that "All swans are white" when we sight a bunch of white swans? There is a serious flaw in this argument. Can you spot it?

The argument is viciously circular because it uses an inductive inference to justify making an inductive inference. It is saying that the whiteness of swans is nonaccidental because of the success of past inductive claims regarding the color of ravens and flamingos. But is this past success nonaccidental? We are still left with the thorny problem of distinguishing between causal and accidental regularities.

Difficulties with induction can also come from other quarters. The statement "All swans are white" seems to be logically equivalent to the statement "All nonwhite objects are not swans." In other words, it seems to be a point of logic that any observation that supports the first statement also supports the second statement, and vice versa. Carl Hemple, a philosopher of science at Princeton University, pointed out something unsettling about the logical equivalence of these two statements. What troubled him?

Hemple realized that pink panthers, yellow submarines, red herrings, Brown University, black magic, and orange juice are all nonwhite nonswans, so that they would count as evidence for the truth of the statement "All swans are white." It is odd indeed that the more pink panthers you see, the more confidence you should have in all swans being white.

All nonwhite objects are not swans.

Do you see a way out of this predicament? Some philosophers have noted that the number of white swans is extremely small compared with the number of nonwhite nonswans. This makes the sighting of a white swan carry far greater weight in supporting the statement "All swans are white" than the sighting of a pink panther. In some crazy way, nonwhite nonswans may support all swans being white, although the support is minimal. The paradox may disappear if we consider the degree of support each kind of sighting provides.

Another way to deactivate the paradox is to dispute its premise: the logical equivalence of "All swans are white" and "All nonwhite objects are not swans." On this premise, the statement "All swans are purple" is logically equivalent to the statement "All nonpurple objects are not swans." Now consider a green thumb. Since a green thumb is a nonwhite nonswan, it is a confirming instance of "All swans are white." By the same token, since a green thumb is a nonpurple nonswan, it is a confirming instance of "All swans are purple." This result is clearly unacceptable. We cannot allow one observation, a green thumb, to support two contradictory conclusions: "All swans are white" and "All swans are purple." The problem must be in the premise of the argument. Therefore, "All swans are white" is not logically equivalent to "All nonwhite objects are not swans."

CHAPTER 10

Lora Does a Möbius Strip

*A mathematician confided
That a Möbius band is one-sided,
And you'll get quite a laugh
If you cut one in half,
For it stays in one piece when divided.*

Anonymous

THE MÖBIUS STRIP is the plaything of recreational topology, by far the sexiest branch of mathematics.

Topology, or rubber sheet geometry, is the study of geometric properties that endure distortion (where distortion does not include ripping, making holes, or filling in holes—except in certain well-defined situations that are too complex to touch on here). Topological shapes are delightful because many of them can be made with just a pair of scissors, paper, and tape. A Möbius strip, or band, can easily be contructed from a long narrow strip of paper, such as adding machine tape. Give the strip a half-twist and then connect the two ends with tape to form a closed ring.

Sure enough, if you cut the Möbius band in half along the band, it remains in one piece, as the limerick promises.

The Möbius band, which was investigated by August Ferdinand Möbius, a nineteenth-century German astronomer and mathematician, has two strange features: It has only one side and it has only one edge. If you run a paint brush along the band, you will find that when you return to where you started, you have painted the entire surface of the band. And if you run a Magic Marker along the edge, you'll soon convince yourself that the band has only one edge.

The Klein bottle is a peculiar topological shape of which only imperfect models can be made. Named after another German mathematician, Felix Klein, the bottle has one side, like the Möbius band, but no edges. The illustration is an imperfect representation because it shows an edge where the narrow neck of

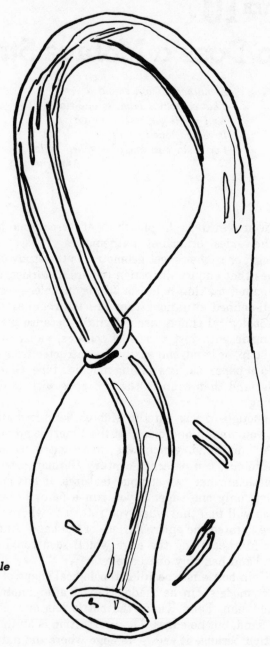

Klein Bottle

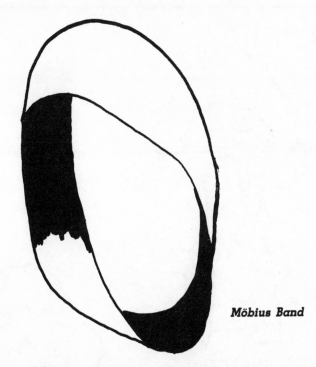

Möbius Band

the bottle penetrates the side. In a painting or a sculpture, there is no way to eliminate this edge. The topologist, whose right-brain hemisphere has a heightened sense of spatial perception, simply imagines that the edge is not there!

Topologists also ponder mundane shapes. What, for example, do a pitcher, the letter A, a car key, a coffee cup, the number 9, and a doughnut have in common? The perceptive paradoxologist knows that each of these objects has a hole. To the mathematician they are "topologicially equivalent": Any of the six objects can, in theory, be transformed into any of the others by being stretched, bent, twisted, shrunk or otherwise deformed without being torn or punctured. Imagine that each object is made of soft rubber; any one of them could be molded into any other.

Stretching without ripping is the basis of the old topological parlor trick of taking off a vest while still wearing a jacket. In the illustration on the next eight pages, Lora does her own version of the trick. She strips off her button-in-the-front bra from under her blouse.

UNBUTTON

PUT LEFT HAND
THROUGH STRAP

3

...PUSH ARM
THROUGH...

4

...PULL STRAP UP
TO SHOULDER...

5

...LIKE SO...

6

...NOW PULL SHIRT
THROUGH LEFT STRAP...

7

...PULL STRAP OVER
TO RIGHT SIDE...

8

...REACH BACK AND GRAB
LEFT STRAP WITH RIGHT HAND...

9

... THREAD RIGHT HAND THROUGH ...

10

...PULL ARM THROUGH...

11

... SLIP STRAP OVER
RIGHT SHIRT COLLAR...

12

THE BRA IS NOW INSIDE
THE SHIRT HANGING FROM
THE RIGHT SHOULDER

13

. . . STUFF BRA INTO
RIGHT SLEEVE . . .

14

. . . GRAB IT WITH
LEFT HAND . . .

15

... PULL IT
OUT...

16

THERE!

Topology is concerned with permanence. The permanent properties of a Möbius band include its one edge and its one side. Another permanent feature of a Möbius band involves the way in which it can be colored. Consider a map of the borders of imaginary countries drawn on a flat sheet of paper. No matter how numerous the countries and how intricate the borders, at most four colors are needed to paint the map in such a way that no two neighboring countries have the same color. (Neighboring countries are defined as those that share a border of more than one point; in other words, two diagonally adjacent squares on a chessboard are not neighbors because their common border has only one point.) This sweeping generalization, called the four-color-map theorem, was proposed in 1852. For the next 124 years the ablest mathematicians tried to prove it, but none succeeded until 1976. The proof made unprecedented use of high-speed computers: 1,200 hours logged on three computers.

A similar map theorem applies to all maps that can be drawn on a Möbius band. I am not going to ask you to prove the theorem. I am merely going to request that you figure out the maximum number of colors required to color any map on a Möbius band in such a way that no adjacent countries have the same color. I also want you to find the maximum number of colors that are needed for any map on a torus, which is topological argot for the surface of a doughnut. The answers are diagrammed on page 142.

Color is also fundamental to Rubik's Cube, or Büvös Kocka, as the Cube is called in Budapest, where it was invented by Ernő Rubik, a teacher of architecture. The $3 \times 3 \times 3$, six-color Cube, of which the Ideal Toy Corporation has sold more than 25 million, has infuriated mathematicians, computer specialists, and puzzle lovers all over the world who have struggled, often in vain, to restore a scrambled Cube to its pristine state. In New York and other cities, street-corner vendors peddle not only the original Magic Cube but knockoff versions, including miniatures incorporated into key chains, pendants, and even tie clasps. Also available is the Royal Cube, with Prince Charles on one side and Lady Di on the opposite side; by judiciously messing up the Cube you can see what Lady Di would look like with the Prince's hair. There is also the Nude Cube, which has Playboylike bunnies on each face; it is particularly difficult to restore the Nude Cube to the virginal state because it is tough to remember which breasts go with which face. Finally, there is the Boob Cube—it is green on all sides so that it can never be scrambled. A Taiwanese imitation of the Magic Cube is rumored to have sold 75 million. As *Time* magazine aptly put it, Búvös Kocka is "the hottest number to come out of Budapest since the Gabor girls went west." In deed, Ideal hired the not-so-cubical Zsa Zsa to promote the Cube.

A few "cube hackers" and "cubers" who aspire to be "cubists," "cubemeisters," and "cubologists" (words from the delightful argot of "cubology," the science of twiddling "cubies"—the Cube's minicubes—into their proper "cubicles") have asked me to reveal the secret of restoring a messed-up Cube to its original state. To be sure, revelation is what is needed because the Cube can be jumbled in 43,252,003,274,489,856,000 distinct ways. Although no one has yet discovered "God's algorithm"—the fewest steps needed to restore the Cube from its most chaotic position—never-fail, step-by-step recipes are known for unscrambling the Cube.

In the definitive cubological work, *Notes on Rubik's Magic Cube* (Enslow Publications, 1981), David Singmaster offers a two-hundred-move solution for restoring the Cube. (The high priests of mathematics think that God's algorithm has at most twenty steps.) Serious Cube mavens should consult the book.

On a lighter note, Singmaster describes how to care for the Cube and relates some cubic lore. To lubricate the Cube, he recommends French chalk and silicone grease. A stiff Cube can be as hard on the body as it is on the mind. Singmaster reports that Dame Kathleen Ollerenshaw, a former lord mayor of Manchester, England, and an amateur mathematician, underwent surgery for "cubist's thumb," a form of tendonitis that is more commonly seen in "teenage disco freaks as a result of too much finger snapping and in overly enthusiastic users of [nonelectric] hedge clippers." And if you start finding the Cube too easy to solve, you might do what John Conway of Cambridge, England, is reported to do: He restores the Cube behind his back, with only four or five "peeks."

Why the Magic Cube has taken the world of mathematics by storm was the subject of Douglas R. Hofstadter's "Metamagical Themas" (an anagram of "mathematical games") in the March 1981 issue of *Scientific American*. (Before you read any further,

Is it possible to achieve any of these positions of Rubik's Cube? (The back cubes are reflections of the front cubes so that the hidden faces can be seen.)

The Quark

The Quark-Antiquark Pair   *Three Quarks*

please try the Cube puzzles on pages 134–5. The discussion that follows gives the answers away.) Hofstadter presents an amazing parallel between magic cubology and particle physics that was discovered by Solomon W. Golomb. More than two hundred kinds of subatomic particles, including the proton and the neutron, are thought to be made up of fundamental entities called quarks. The quark has an electric charge of $+\frac{1}{3}$ and its antimatter sibling—the antiquark—has a charge of $-\frac{1}{3}$. Neither the quark nor the antiquark has ever been observed in isolation. They have been detected inside particles and then only in three configurations: as a quark-antiquark pair, as three quarks, or as three antiquarks.

In cubology it is impossible to twiddle the cubies so as to achieve a state in which a single corner cubie is twisted a third of a full turn in either direction while the rest of the Cube is in order. A clockwise one-third twist of a cornie cubie is called a quark (remember: the particle quark has a charge of $+\frac{1}{3}$); a counterclockwise one-third twist of a corner cubie is called an antiquark. Thus a cubie quark, like a quark in particle physics, cannot be seen in isolation.

The parallel is even deeper. The observable quark-antiquark pair in particle physics has an analogue in the Magic Cube: A Cube state can be achieved in which all the cubies are pristine except for one cubie quark and one cubie antiquark. Moreover, three quarks can be observed in particle physics, and so can three cubie quarks. The parallel extends to three antiquarks in cubology.

What is the significance of the parallel? "In the cubical world," Hofstadter concludes, "the underlying reason for quark confinement lies in group theory. There may be a closely related group-theoretical explanation for the confinement of quark particles. That remains to be seen, but in any event the parallel is provocative and pleasing."

The Möbius band can also serve as a model of phenomena in particle physics. The one-sidedness of a Möbius band is an analogue of how the rotation of a subatomic particle, such as a neutron, will affect its spin.

When a spinning top is rotated 360 degrees, it continues to spin in the same direction in which it was originally spinning. Nevertheless, when a spinning subatomic particle is rotated 360 degrees, it will be spinning in the opposite direction. Another 360-degree rotation (or a total rotation of 720 degrees) is needed to restore the particle to its original spin state. This counterintuitive effect has an analogue in the Möbius band: An object moving along the surface of a Möbius band must make not one full circuit (a 360-degree rotation) but two full circuits (a 720-degree rotation) in order to return to the point at which it started. Indeed, one circuit will put the object opposite its starting point, as one rotation of a spinning subatomic particle will make it spin in the opposite direction.

The Möbius band is of service not only to recreational mathematicians and particle physicists but also to industrialists. The B. F. Goodrich Company, for example, has a patent on a conveyer belt made from a Möbius band. One side of a conventional conveyer belt is subject to more stress than the other. In a Möbius belt, however, the wear and tear is spread out over "both sides," so that the belt lasts twice as long.

Puzzles

I have five topology problems for you. I suggest that you approach each problem by thinking about it in the abstract and then testing your answer by constructing the relevant topological shapes.

1. Make a Möbius band and cut in it half around the middle. What will happen if you cut the resulting shape in half?

2. Make a Möbius band and imagine cutting it around its girth a third of the way from the edge (see diagram A). The cut will take you twice around the band before you return to where you started. What shape do you end up with?

3. Take a strip of paper and cut four slits in it, as shown in diagram B. Now tape end 3 to end 6. Pass end 2 under end 3 and tape it to end 5. Join end 1 to end 4 after weaving end 1 over end 2 and under end 3 and after weaving end 4 over end 6 and under end 5. What will happen if you complete the cuts around the band? This problem comes from *Mathematical Diversions*, by J. A. H. Hunter and Joseph S. Madachy, Dover, 1975.

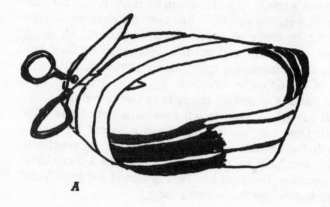

A

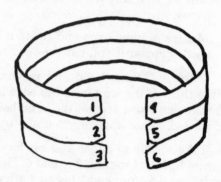

B

4. Imagine a Möbius band sixteen inches around and of infinitesimal thickness in which two R-shaped organisms of infinitesimal thickness are embedded. All I mean by infinitesimal thickness is that you can see the R-shaped organisms through the band. A good way to construct a model of what I have in mind is to make the Möbius band from a thin, flexible piece of transparent plastic (tracing paper will also do). The organisms can be made by taking a circular piece of transparent plastic the size of a penny and drawing an R on it. When the R is held against the Möbius band, it is easy to imagine that the R is embedded in the strip.

The R-shaped organisms are restricted in their movement. They cannot leave the strip because they are embedded in it.

Two organisms were born at the same time in the configuration RR across the width of the strip.

"I'm taller than you" were the first words uttered by one of the Rs.

"No, I'm taller," insisted the other R.

"We'll see about that. Let's compare height the way earthlings do, by standing back to back."

No matter how they tried, however, they could not orient themselves as ЯR. Can you suggest a way?

This problem applies not only to R-shaped organisms but to any asymmetrical creatures embedded in a Möbius band. For example, can the creatures shown in diagram C arrange themselves as shown in diagram D?

5. Take two identical 16-inch-long strips of transparent plastic and put them on top of each other. Now give the sandwich of strips a half-twist and join the ends to create two nested Möbius bands. To convince yourself that nothing funny happened when the pieces were joined, insert a pencil between the bands and move it so that it makes a complete circuit between them; sure enough, you're dealing with two unconnected Möbius bands.

Again, imagine two R-shaped organisms embedded in the bottom band, born as before in the configuration RR. Can these organisms also orient themselves back to back? "You've gotta be kidding," I hear you cry. "You've just added the second band to befuddle me, but the situation is exactly the same as the previous case." Is it really the same? Take a good hard look.

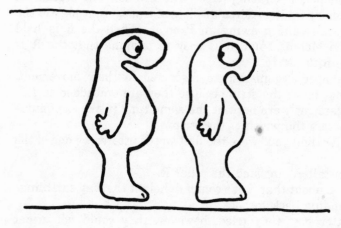

C

D

ANSWERS

1. The result is two interlocked loops, each of which has four half-twists.

2. The result is two interlocked bands: a mundane two-sided hoop and a Möbius band.

3. The result is three interlocked rings.

4. Either organism should move forward sixteen inches, making a complete circuit around the band. When the organism rejoins its twin, they can easily be oriented as mirror images art R. If the organisms were embedded not in a Möbius band but in a two-sided loop, they could never stand back to back.

5. The situation is entirely different. The R-shaped organisms can never orient themselves back to back. If either organism moves forward sixteen inches to make a complete circuit, as in the previous problem, it will end up on the top band, above its twin on the bottom band. And if the organism moved forward another sixteen inches, it would return to its original position on the bottom band, and the organisms would have their original orientation, RR. How, you may wonder, can the organism switch bands?

Leaping or flying from one band to the other would scarcely qualify as staying embedded in the band. What's the trick? Unravel the band and the secret will reveal itself. There are not two bands but one. And it is not a Möbius band. It has two sides and two edges. The band could be formed by taking a narrow strip of paper and giving it four half-twists before joining the ends. (I am indebted to Martin Gardner, whose writing first acquainted me with the paradoxical nested band.)

This problem and the previous one bring out an important feature about the construction of Möbius bands. If a strip is given an odd number of half-twists before its ends are joined, the result is a one-sided, one-edged band; in other words, a Möbius band. If a strip is given an even number of half-twists before its ends are joined, the result is a two-sided, two-edged band. Two asymmetrical organisms lined up in a row across a band can orient themselves as mirror images only if the band is a Möbius band.

Six-color-map theorem for a Möbius band
Join A to D and B to C to make a Möbius band.

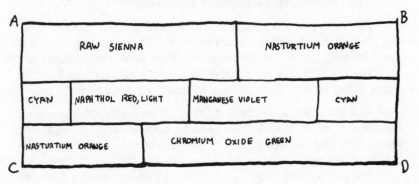

Seven-color-map theorem for torus
Join A to C and B to D. Then bend the tube into a torus.

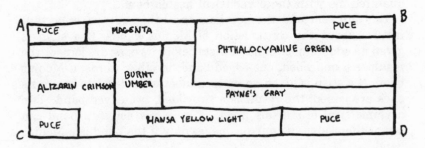

CHAPTER 11

The BANG and Other Phrenetic Phrases

I N THE MAY 1981 issue of *Science Digest* Lora and I pointed out that the physical arrangement of letters can communicate a phrase. For example, the big letters "BANG" in the upper box suggest "Big bang" and the letters in the lower box represent "a big fish in a little pond" because the word "FISH," which is big, is in the word "pond," which is little. On the next four pages are forty-eight clever phrenetic phrases, culled from my mind and from 1,500 letters sent to me by readers. The best phrases were from Bill Dembowski, Johnstown, PA (no. 29) and Dennis M. Foreht, Weston, Ontario (nos. 35, 37, 42, and 44).

FA*π*CE 1	BAKED 2	SUNNY CLOUDY WINDY I AM 3
NCLE SAM 4	DEKCAB 5	STOMACH 6
EYE LOO EMK ME 7	PHD BA MD 212 8	A N C E R 9
MILK 10	SU144LT 11	 12

CHANCE

1. Pie in the face
2. Half-baked
3. I am under the weather
4. Uncle Sam needs you
5. Backed into a corner
6. Stomach cramp
7. Look me square in the eye
8. Three degrees above boiling
9. Topless dancer
10. Evaporated milk
11. A gross insult
12. Outside chance

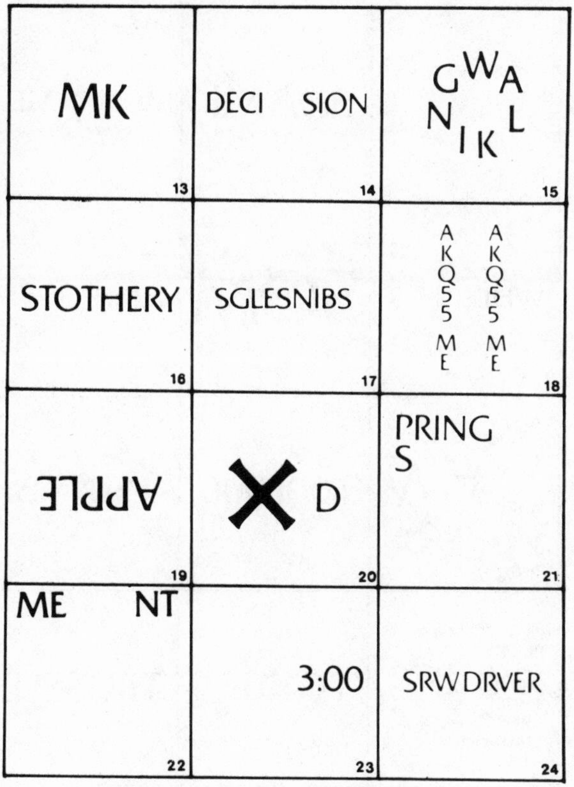

MK 13	DECI SION 14	G W A N I K L 15
STOTHERY 16	SGLESNIBS 17	A A K K Q Q 5 5 5 5 M M E E 18
APPLE 19	**X** D 20	PRING S 21
ME NT 22	3:00 23	SRW DRVER 24

13. Condensed milk
14. Split decision
15. Walking in a circle
16. The inside story
17. Mixed blessings
18. Hand-me-downs
19. Apple turnover
20. Cross-eyed
21. Spring is just around the corner
22. Penthouse apartment
23. The time is right
24. Screwdriver without ice

| | IMAGE | MD   MD |
| 25 | 26 | 27 |

| WHEATHER | SOM | LAWPORKY |
| 28 | 29 | 30 |

| I R A N E A B D (scattered) | 1,001,000 | PANTS |
| 31 | 32 | 33 |

| PIE TORTE / PASTRY BREAD / CAKE CUPCAKE / MUFFIN BETTY / TART MERINGUE / CRISP COOKIE / STRUDEL | (mirrored letters) | TRUSSLE |
| 34 | 35 | 36 |

25. Two albino hippopotami with white white silk stockings making love in a pile of snow under cloudy skies while a flock of doves watches closely and laughs
26. Mirror image
27. Paradox (by Glenn Brock, Austin, MN)
28. A bad spell of weather (by Bob and Claudine)
29. This could be the start of something big
30. Mixing work and play
31. Scatterbrained
32. One in a million
33. Fancy pants (by Delores Moffeit and Betty-Jo Hartman, Runnemede, NJ)
34. Baker's dozen
35. Tooth decay
36. Mixed results

⅃ᴎᴚ˅ 37	P I 38	SCN 39
X X X X X DEC.X X 31 X X X X X 40	(scribbles) 41	HN AEYSE TDALCEK 42
SOTA 43	SLOW SNOW STOW 44	*ure* 45
NE IRS 46	WHITE 47	WOOD DALE MASSACHUSETTS 48

37. Slipped disk
38. Bottomless pit (by Les Greene, Fresno, CA)
39. *Science Digest* (by Don Boyette, Memphis, TN)
40. New Year's Eve in Times Square (by Darrell Smith, Portland, OR)
41. Three's a crowd (by Welles Price, Winter Haven, FL)
42. Needle in a haystack
43. Minnesota (by Louis L. Atkin, El Paso, TX)
44. No show
45. Scripture
46. Half mine, half yours (by Lisa Housman, age 9, Watertown, MA)
47. White lies
48. Dale Underwood, Andover, Massachusetts (by Kent Gray-bill, Spokane, WA)

CHAPTER **12**

How's Your Resistance?
(There's No Place like Ohm)

AS A PLAYFUL PARODY of my phrenetic phrases, Keith Weber and his co-workers at the Southern Texas Project Electric Generating Station in Bay City, Texas, sent me thirteen phrases that involve the word "ohm." The last nine phrases are mine.

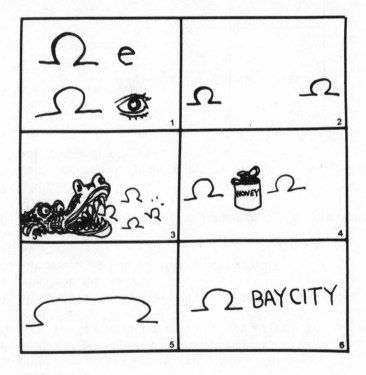

1. Ohm me ohm my
2. Ohm away from ohm
3. Ohm eater
4. Ohm sweet ohm
5. Ohm stretch
6. Ohm town

15. Ohm sick
16. Ohm wrecker
17. Down ohm
18. Ohmar Sharif
19. Ohming pigeons
20. When in Ohm do like an Ohman
21. Ohmley
22. Ohmer

7. Ohm on the range
8. Mobile ohm
9. Ohm work
10. Ohm plate
11. Ohm rule
12. Ohm run
13. Designer ohms
14. Broken ohm

CHAPTER 14
Triskaidekaphobia

THE FORCES OF EVIL did it again. They conspired to make the thirteenth of August 1982 fall on a Friday. In 1981 they even went so far as to serve up three Friday the thirteenths, in February, March, and November. Triskaidekaphobia, morbid fear of the number 13, is a phobia that afflicts many cultures. Is it mere superstition, or is there good reason to think that the number 13 is endowed with supernatural power? Indeed, overwhelming historical and mathematical evidence links it with evil and misfortune. After all, Jesus was the thirteenth at the Last Supper and look what happened to him.

The conviction that disaster stalks a group of thirteen diners at a table goes back even further than the Last Supper. Nordic mythology has it that the evil god Loki, the chief author of human misfortune, crashed a banquet of other deities, swelling their ranks to thirteen. After hurling insults at the gathered company and getting himself thrown out, Loki arranged for the death of one of the diners, the god Balder.

The fear of seating thirteen at a dining table still thrives today, although it is probably only remotely influenced by the knowledge of Loki's misadventures or reminders of the Last Supper. In France a *quatorzième,* or professional fourteenth guest, can be hired on the spur of the moment to round out a dinner party.

In his delightful article "Memorable Meals: Thirteen at Table" in *Gourmet* magazine, Vincent Starrett cites stories of luminaries, such as Herbert Hoover and Franklin Delano Roosevelt, who were plagued by the superstition. Grace Tully, who was

F.D.R.'s secretary, wrote of Roosevelt: "The Boss was super-stitious, particularly about the number thirteen and the practice of lighting three cigarettes on a single match. On several occasions I received last-minute summonses to attend a lunch or dinner party because a belated default or a late addition had brought the guest list to thirteen. My first invitation to a Cuff Links Club dinner, held annually on the President's birthday, came about in 1932 when withdrawal of one of the guests left a party of thirteen. I was an annual fixture after that."

The funniest anecdote Starrett turned up concerns the Thirteen Club in London, which was started in 1890, six years after the founding of the Thirteen Club in New York. The total number of members, the number of members who dined together at a table, the membership dues, and the dates of the meetings all involved the number 13 or a multiple of it. The autobiography of Harry Furniss, a member of the London Thirteen Club who drew caricatures for the London humor magazine *Punch,* contains a letter he received in 1894 from Christiana, Norway:

"Sir—I see you are going to have an anniversary dinner on

the 13th of this month, and I take the liberty to send you the following: In 1873, March 20th, I left Liverpool in the steamship *Atlantic*, then bound for New York. On the 13th day, the 1st of April, we went on the rocks near Halifax, Nova Scotia. Out of nearly 1,000 human beings, 580 died.

"The first day out from Liverpool some ladies at my table discovered that we were thirteen, and in their consternation requested their gentleman-companion to move to another table. Out of the entire thirteen, I was the only one who was saved. I was asked at the time if I did not believe in the unlucky number 13. I told them I did not. In this case the believers were all lost and the unbeliever saved. (Signed) N. Brandt."

So great is the power of the number 13 that it can even undermine the progress of science. Although most architects have enough humility to bypass the number 13 when labeling the rooms and floors of a building, the National Aeronautics and Space Administration takes no such precautions—and it has suffered for it. In April 1970, NASA haughtily launched the thirteenth Apollo mission, at none other than 1313 hours central time, from pad 39 (39, of course, is the third multiple of 13). The astronauts, whose first names (James, Fred, and John) contain a total of thirteen letters, were scheduled for rest periods that began at thirteen minutes past the hour. The dark forces do not like to be toyed with; on April 13 an oxygen tank exploded and the mission had to be aborted.

NASA is a slow learner. In November 1981, it was forced to cut short the second flight of the Space Shuttle because of a glitch that developed in the fuel cells on Friday the thirteenth.

The diabolical power of the number 13 has been known to torment men of otherwise rational genius. A classic example is the German composer Richard Wagner, whose brilliant operas drew heavily on myths and legends involving unnatural death, wholesale destruction, and assorted other calamities. The sinister forces that surfaced in his music undoubtedly stem from the many 13s in his life.

Richard Wagner, whose name has thirteen letters, was born in 1813 and died on February 13, 1883. When he was thirteen and attending school in Dresden, his mother and sisters moved to Prague, leaving the miserable boy behind. At the age of twenty-six (which is two times 13) bad luck struck again; he had incurred such huge debts that he was forced to flee to Paris to evade his creditors. Of Wagner's major operatic works, only thirteen survive in complete form; he destroyed large parts of his first opera, which was premiered on February 13, 1838. The first performance of his *Tannhäuser*, whose score was completed on April 13, 1845, was not well received; and the revised version that opened in Paris on March 13, 1861, was an unqualified critical disaster.

The first complete performance of Wagner's huge, gloomy cycle, *Der Ring des Nibelungen*, in which the hero is overcome by dark forces, was begun on August 13, 1876. Wagner scored the last bars of *Parsifal* on January 13, 1882, and September 13 of that year was the last day he spent in Bayreuth, where his operas are annually performed in summer festivals.

Evidence for the power of the number 13 comes from many quarters, often through the link of one person with twelve associates. Evil covens were traditionally formed by twelve witches together with a devil. The pervasiveness of the number can also be seen in King Arthur and his twelve chief knights, Jacob and his twelve sons, Odysseus and his twelve companions on the island of Cyclopes, and Romulus and the twelve vultures that legitimized him as the founder of Rome. When Merlin vanished from the earth, he took with him thirteen treasures that had magical qualities. And on a modern note, the Cuban missile crisis lasted thirteen days.

Wait a moment, I hear you cry, two can play this game. There's luck in the number 13 if you look in the right places. The tormented Wagner may have been born in a year ending in thirteen, but the good Thomas Jefferson was born on the thirteenth of April 1743. It was thirteen states that formed the glorious union, and indeed the motto *E Pluribus Unum* has thirteen letters. Justice itself recognizes the virtues of thirteen in a judge and twelve jurors.

Lora shoved under my nose her idea of an amusing poem, "Lucky Thirteen" by Priscilla Oakes, which attempts to counter the bad reputation of the baker's dozen. It begins:

> *That 13 brings ill-luck is mere tradition*
> *Yea, even more, 'tis foolish superstition—*
> *For 13, lucky number, runs,*
> *Through all our Country's ripening suns.*

What the poem lacks in wit and meter it more than makes up for in enthusiasm. For thirty-two pages, "Lucky Thirteen" goes through American history and describes all the good things that have happened on the thirteenth of the month. The poem, which was written in 1919, just after World War I, is followed by a list of 182 "lucky" phrases, each composed of thirteen letters. If this contrived list is the most that can be said for the goodness of 13, then the number is certainly diabolical. It is a puzzle in itself to decipher the meaning of some of the phrases, which include "Germany Guilty," "American Women," "Secret Service," "Diamond Brooch," "Huge Horsehoe," "Invest in Bonds," "Booming of Guns," "Stopped the Hun," "A Divine Moment," "Sorbonne Guest," "His Abdication," and "Home Sweet Home."

I turn now to the mathematical evidence for the far-reaching power of the number 13, which I offer in the form of two problems for you to solve.

1. Presumably, the dark forces would not want a year to go by without having a day to wreak their havoc on the world. Prove, then, that there is at least one Friday the thirteenth every year. What is the maximum number of Friday the thirteenths that the powers of evil can serve up in a given year? What maximum-dose years are left in the current century?

2. On what day of the week is the thirteenth of the month most likely to fall? You can pick up a clue to the solution of this problem if you were clever enough to notice that I gave an incomplete analysis in "Sleight of Mind in Katmandu" of the frequency with which the evil spirits conspire to make Friday fall on the thirteenth of two consecutive months. I worked out the pattern up to the year 2036, and concluded that starting in 1981 the number of years between years with back-to-back Friday the thirteenths follows the pattern 6, 11, 11, 6, 11, 11. My analysis is airtight as far as it goes but there is more to the picture. (Lora attributed my lapse to triskaidekaphobia, but I prefer to think of it as being due to her distracting behavior while I was making the calculations.)

ANSWERS

1. The solution I offer is based on an elegant proof by William T. Bailey in *Mathematics Teacher* (1969, pp. 363–64). Whether the thirteenth of a given month falls on a Friday depends on two

things: the day on which the thirteenth fell in the previous month and the number of days in the previous month. For example, if the thirteenth of January was a Tuesday, the thirteenth of February will be a Friday (three days later) because January has thirty-one (twenty-eight plus three) days. Now look at the table below. Column A, starting with December, lists the months in consecutive order. In column B, the first list gives the number of days in each month in a nonleap year. Next to it is the number of days the thirteenth will move forward from that month to the next. Finally, there is a list of the number of days the thirteenth of each month will move forward with respect to the day on which the thirteenth fell the previous December. Because this list includes all the possible shifts—from no days to six days—you can see that the thirteenth falls on each day of the week at least once in every nonleap year. Thus not one nonleap year can pass without a Friday the thirteenth. Of all the numbers, 6 appears the most in the third list—namely thrice—and so in any nonleap year the thirteenth can fall at most three times on a given day of the week. Hence three Friday the thirteenths is the maximum dose of evil that the sinister forces can serve up in a nonleap year, as they did in 1981. This triple whammy occurs whenever the thirteenth of the previous December is a Saturday.

Column C gives the same information calculated for a leap year. The result turns out to be the same: There is at least one

A	B Nonleap Year		C Leap Year	
December	31 3	3	31 3	3
January	31 3	6	31 3	6
February	28 0	6	29 1	0
March	31 3	2	31 3	3
April	30 2	4	30 2	5
May	31 3	0	31 3	1
June	30 2	2	30 2	3
July	31 3	5	31 3	6
August	31 3	1	31 3	2
September	30 2	3	30 2	4
October	31 3	6	31 3	0
November	30 2	1	30 2	2

Friday the thirteenth and at most three. There are three whenever the thirteenth of the previous December is a Tuesday.

There are only three triple-whammy years remaining in this century: 1984, 1987, and 1998.

2. In my analysis of back-to-back Friday the thirteenths, which can only occur in nonleap years, I failed to mention that not every year whose number is divisible by four is a leap year. If the tropical year were precisely 365 days 6 hours, it would make sense to have a leap year every four years so that the tropical year and the calendrical year would remain consistent. The tropical year, however, falls short by 11 minutes 14 seconds; it is only 365 days 5 hours 48 minutes 46 seconds. With a leap year every four years, our calendrical year would be out of step by one day every 128 years, or approximately three days every 400 years. Therefore, in every 400-year period three years that would otherwise be leaped should not be leaped. By convention, these three years are those whose number is divisible by 100 but not by 400. In other words, the year 1700, 1800, 1900, 2100, 2200, and 2300 are not leap years, whereas 1600 and 2000 are leap years.

What all this means is that it takes a full 400 years for our calendar to make a complete cycle and start over, so that any 400-year period contains the same number of occurrences of a date on a certain day of the week. The pattern I gave for back-to-back Friday the thirteenths holds only until the year 2100, which is the first unexpected nonleap year.

To determine the day of the week on which the thirteenth is most likely to fall, one must examine any complete 400-year period. I find that the thirteenth is on Sunday 687 times, on Monday and Tuesday 685 times each, on Wednesday 687 times, on Thursday 684 times, on Friday 688 times, and on Saturday 684 times. You guessed it: The thirteenth is most likely to fall on Friday. I see this as the strongest evidence in favor of the power of Friday the thirteenth. One would expect the diabolical number 13 to designate as evil the day of the week on which the thirteenth was most likely to fall. An elegant paper, "To Prove That The 13th Day of the Month Is More Likely To Be a Friday Than Any Other Day of The Week," appeared in the *Mathematical Gazette* (1969, pp. 127–29). It was written by a student, S. R. Baxter, aged thirteen.

CHAPTER 15

Freud and the Number 23

THE STRANGEST LETTER I received about my musings on triskaidekaphobia had no return address. I want to quote it in full.

Dear Dr. Krypton,

I was fascinated by your comments on the number 13. Perhaps the power of the number stems in part from the completeness of its predecessor, the number 12, which manifests itself in the twelve months of the year, the twelve signs of the Zodiac, the twelve hours of the day, the twelve gods of Olympus, the twelve labors of Hercules, the twelve tribes of Israel, the twelve apostles of Jesus, the twelve days of Christmas, and the twelve eggs in a carton of eggs. In other words, the number 13 is restless to the point of being terribly powerful because the number is jealous, having barely missed the tranquil harmony that comes with being the number 12. In numerology the number 22 is also complete. After all, there are twenty-two letters in the Hebrew alphabet and twenty-two major trumps in a deck of Tarot cards. If my hypothesis about the origin of the power of the number 13 is correct, then the number 23 should be powerful too. But I don't know whether it is. That's why I'm writing to you. I hope that in your infinite wisdom you can shed some light on this matter.

Cryptically yours,
MR. PTYLMXKZ

Let me first say that this letter is representative of an increasing but disturbing tendency of readers to call me Dr. Krypton. I know my conundrums and penetrating insights may torment people to the point where they feel they've lost all power, but my name is not derived from kryptonite, the one substance that strips Superman of his power. The name comes from my parents, Arnold and Sadie Crypton. I am also taken aback by correspondents, who have bestowed on me, however affectionately, the nickname Crypt.

This letter, however, not only has a puzzling greeting but also a baffling closing. "Cryptically yours, MR. PTYLMXKZ." It registered on Lora's perceptive mind that the name contained no vowels. She suspected that the name was in code because it seemed too difficult to pronounce for it to be a pseudonym.

I wondered whether there might be a connection between the weird signature and the misspelling of my name. Sure enough, I found one. My computerlike brain quickly discerned that MR. PTYLMXKZ is an anagram of MR. MXYZPTLK, Superman's foe from the fifth dimension. Having failed to defeat Superman, MR. MXYZPTLK has turned his diabolical attention on me. I realized it was critical I meet his challenge of providing historical and mathematical evidence for the number 23, as I had done for the number 13.

Lora found a good place to start. She reminded me of a *Science Digest* article (January, 1982) that described many significant 23s in the life of novelist William Burroughs. In 1958 a Captain Clark told Burroughs that he had been sailing for twenty-three years without incident. Later in the day Clark had an accident. That evening Burroughs was pondering the untimely accident when he heard a radio bulletin describing the crash of an Eastern Airlines plane, Flight 23, piloted by another Captain Clark.

Arthur Koestler has a weakness for numerology, and his book *The Challenge of Chance* (Random House, New York, 1973) includes a letter from Hans Zeisel, a professor at the University of Chicago Law School, whose life was plagued by the number 23. The letter was dated August 23. Zeisel had lived in Vienna at Rossauerlände 23 and had worked in a law office at Gonzagagasse 23. His mother had lived at Alserstrasse 23 and her parents at Gablonz Mozartgasse 23. One spring his mother traveled to Monte Carlo, where she happened to start reading Ilya Ehrenburg's *The Love of Jeanne Ney*. A character in the book wins money at roulette on the number 23. That convinced Zeisel's mother that she should bet money at roulette on the number that had haunted her family. Sure enough, 23 came up on the second try.

Such anecdotes are pale in comparison with the story of Sigmund Freud's seven-year flirtation with the number 23. The man who did the most for the number 23 is Wilhelm Fliess, a Berlin nose-and-throat surgeon who from 1893 to 1900 was Freud's closest friend and intellectual companion. In his 1897 monograph *The Relations Between the Nose and the Female Sex Organs from the Biological Aspect*, Fliess maintained that every person has both feminine and masculine traits that manifest themselves according to certain numerological laws. The feminine characteristics follow a twenty-eight-day cycle, as menstruation does. The masculine cycle is twenty-three days, presumably because there are roughly that many days between the end of one menstrual period and the start of the next. (Triskaidekaphobes must not forget that women are plagued by the curse thirteen times a year.) The two cycles, which are transmitted from mother to child, govern birth, illness, and death. Moreover, Fliess wrote, "The wonderful accuracy with which the period of twenty-three, or, as the case may be, twenty-eight whole days is observed permits one to suspect a deeper connection between astronomical relations and the creation of organisms." Indeed, the lunar cycle is twenty-eight days.

In what is surely the strangest chapter in Freud's life, he was much taken by Fliess's fanciful ideas on bisexuality, including the twenty-eight-day feminine cycle and the twenty-three-day masculine cycle. Freud sent Fliess examples of significant 23s and 28s in his own life and the lives of his family and friends.

Freud had distinguished between two kinds of neurosis, and he tried to connect them to the twenty-eight-day and twenty-three-day cycles. Moreover, he suggested that for a person of either sex the release of excess 23 material was pleasurable, whereas that of excess 28 material was unpleasurable.

Freud believed that all known forms of contraception caused anxiety that detracted from sexual satisfaction. Indeed, he reportedly told one of his patients that "the greatest invention some benefactor can give mankind is a form of contraception which doesn't induce neurosis." Freud thought that on the basis of the twenty-eight-day and twenty-three-day cycles, Fliess would be able to determine the days in the menstrual cycle when a couple could engage in sexual congress without having to guard against conception. According to Freud's biographer, Ernest Jones, as early as July 10, 1893, Freud "set his hopes on Fliess's solving the problem 'as on the Messiah,' and a little later (December 11) he promised him a statue in the Tiergarten in Berlin when he succeeded. Two years later (May 25, 1895) it looked as if success were in sight, and he wrote: 'I could have shouted with joy at your news. If you really have solved the problem of conception I will ask you what sort of marble would best please you.'"

Although Fliess was not forthcoming with the days of the menstrual period on which a woman is infertile, Freud expressed even more faith in Fliess's program. On July 12, 1897, Freud wrote to him: "At Aussee I know a wonderful wood full of ferns and mushrooms, where you shall reveal to me the secrets of the world of lower animals and the world of children. I am agape as never before for what you have to say—and I hope that the world will not hear it before me, and that instead of a short article you will give us within a year a small book which reveals organic secrets in periods of twenty-eight and twenty-three." A year later, after Fliess linked the masculine and feminine cycles to periodic phenomena in the cosmos, Freud dubbed hm "the Kepler of biology."

The hero-worship continued in a letter dated August 26, 1898. "Yesterday the good news reached me that the enigmas of the world and of life were beginning to yield an answer, news of a successful result of thought such as no dream could excel. Whether the path to the final goal, to which you desire to use mathematical points, will prove to be short or long, I feel sure it is open to you."

For Fliess's part, he sometimes responded to Freud's letters with a numerological analysis of Freud's life. He warned Freud that fifty-one would be a tough age for him because the sum of 28 and 23 is 51. Freud was afraid. His fear even comes through in his classic work *The Interpretation of Dreams.* "Fifty-one is the age," Freud wrote, "which seems to be a particularly dangerous one to men. I have known colleagues who have died suddenly at that age, and amongst them one who, after long delays, had been appointed to a professorship only a few days before his death."

Fliess came increasingly to see the world in terms of 23s and 28s. When he couldn't explain the date of a significant event in terms of multiples of 23s and 28s, he would explain them in terms of the sum or difference of 28 and 23. He even introduced a seven-and-a-half-year cycle that governed the lives of great men. Freud followed him to a point. On October 11, 1899, Freud reminded Fliess: "According to an earlier calculation of yours 1900–1901 ought to be a productive period for me (every seven and a half years)." On January 8, 1900, Freud chided Fliess for having waited as long as fourteen days to respond to his last letter; what apparently aggravated Freud was that 14 is half of 28.

The two men soon parted company, largely because of Fliess's own psychopathology. He felt that Freud wasn't sufficiently enthusiastic in endorsing the periodic laws of 28 and 23. Moreover, Freud and Fliess had a heated priority fight over the discovery of universal bisexuality. Freud had forgotten that it was Fliess who had taught him that all human beings have a bisexual disposition. Jones notes that Freud's memory lapse was "a very severe case of amnesia. Only a year before he had written: 'You are certainly right about bisexuality. I am also getting used to regarding every sexual act as one between four individuals.'" In any event, their falling out was inevitable because the rigid determinism of Fliess's numerological laws was at odds with the dynamic nature of Freud's theory of behavior.

In 1905 Fliess published his chief work, *The Rhythm of Life: Foundations of an Exact Biology,* which has served as a bible for a generation of gullible people engrossed in biorhythms. In the book, whose second edition was released appropriately enough in 1923, Fliess set out to explain critical dates in terms of the formula $(23 \times A) + (28 \times B)$, where A and B are any integers. As Martin Gardner pointed out in a delightful piece on Fliess in *Scientific American,* Fliess apparently did not realize that the formula $(Q \times A) + (P \times B)$, where Q and P are not restricted to 23 and 28 but are open to any pair of positive integers that have no common divisor, will serve to express *every* positive integer. In other words, the formula $(21 \times A) + (19 \times B)$ yields every positive integer, as well as the formula $(23 \times A) + (28 \times B)$ does.

Freud himself caught on belatedly to Fliess's numerological Ouija game. Jones once asked Freud how Fliess would explain a respite between attacks of appendicitis if the respite was neither a multiple of 23 nor a multiple of 28. Freud replied: "That wouldn't have bothered Fliess. He was an expert mathematician, and by multiplying 23 and 28 by the difference between them and adding or subtracting the results, or by even more complicated arithmetic, he would always arrive at the number he wanted."

Fliess, however, had the last laugh. He died in the twenty-eighth year of this century. And Freud, although he defied Fliess's forecast by living well beyond the age of fifty-one, died on the twenty-third of a month, the same day on which his father had died.

The strongest mathematical evidence for the power of 23 is the role the number plays in the paradox of shared birthdays. Surely you have been in a situation in which a small group of people compared birthdays and found, to their surprise, that at least two of them were born on the same day of the same month. Suppose there are ten people in the group. Intuition may suggest that the odds of two sharing a birthday are quite low. Probability theory, however, shows otherwise: The odds are better than one in nine.

For a group of twenty-two people the odds are slightly against two of them sharing a birthday. When the size of the group reaches twenty-three, however, the odds swing in favor of a shared birthday.

Warren Weaver, the celebrated mathematician who wrote *Lady Luck: The Theory of Probability,* once explained these odds at a dinner party of army and navy officers. The group included twenty-two people, and one proposed that they test Weaver's explanation. "We got all around the table without a duplicate birthday," Weaver recalled. "At which point a waitress remarked, 'Excuse me. But I am the twenty-third person in the room, and my birthday is May seventeen, just like the general's over there.'"

What is the most compelling evidence for the power of the number 23? Why, the twenty-three-letter code word AUFICQ-DBKLTPZYXMSRHOGNE. The meaning of the word will become clear if you pronounce it out loud. Go ahead, try it. Don't be shy. Trust me.

Why did I have you pronounce this ridiculous word? Well, embedded in the word is MR. MXYZPTLK's name spelled backward. Superman legend has it that MR. MXYZPTLK will be propelled to the fifth dimension and confined there for ninety days if he can be made to say his name backward. I trust Lora and I have earned three months of peace. Twenty-three skiddoo, MR. MXY-ZPTLK. Twenty-three skiddoo.

CHAPTER 16

The Puzzles of Nympholepts: Lewis Carroll and Vladimir Nabokov

THIS YEAR IS THE sesquicentennial of the birth of Charles Lutwidge Dodgson—writer, artist, photographer, clergyman, mathematician, parodoxologist, and nympholept. Most of us know him only by his nom de plume, Lewis Carroll. As the author of the *Alice* books, he brought phantasmagorical nonsense verse into houses, huts, and igloos all over the world. *Alice in Wonderland* has been published in at least forty-seven languages, including Swahili, Afrikaans, Gaelic, Oriya, Kanarese, Thai, Esperanto, Braille, and Pitman shorthand.

The oldest of eleven children, eight of them girls, Carroll was an enigma. Although he was a wonderfully zany writer, he was apparently a boring conversationalist. He was fascinated by epilepsy, and he wrote polemics against hunting, vivisection, and the conversion of parks into cricket grounds. An avid theatergoer, he objected to profanity on the stage. For twenty-one years he wrote only with purple ink. For his entire adult life he kept an index of all the letters he sent and received; it contained 98,000 items.

Carroll was fond of little girls. He cavorted with them on the beach, invented puzzles and games for them, and photographed them in the semi-nude. Carroll wrote to the illustrator Harry Furniss: "I wish I dared dispose with all costume: naked children are so perfectly pure and lovely, but Mrs. Grundy would be furious—it would never do. Then the question is, how little dress would content her." In another letter he said: "I confess I do not admire naked boys in pictures. They always seem to need clothes—whereas one hardly sees why the lovely forms of girls should ever be covered up." Indeed, Carroll loathed little boys.

After he refused once to see a friend's son, Carroll wrote: "He thought I doted on all children but I am not omnivorous like a pig. I pick and choose."

In 1877 Carroll invented the now-famous game of Doublets, or Word Golf, for two of his little-girl friends who "had nothing to do." The rules of Doublets were published in 1879 in the magazine *Vanity Fair*. The idea of Doublets is to change one word into another by a series of linking words, each of which differs from the preceding word by one letter only. Carroll noted that *head* could be converted into *tail* by the linking words *heal, teal, tell,* and *tall*. The chain of words is usually written vertically:

HEAD
HEAL
TEAL
TELL
TALL
TAIL

Not all words are appropriate for Doublets. "It is, perhaps, needless to state," Carroll wrote, "that it is *de rigueur* that the links should be English words, such as might be used in good society."

In *Vanity Fair* Carroll posed the following Doublets, for each of which the number of linking words is indicated in parentheses:

1. Drive PIG into STY (4)
2. Make WHEAT into BREAD (6)
3. Dip PEN into INK (5)
4. Change TEARS into SMILE (5)
5. Cover EVE with LID (3)
6. Turn POOR into RICH (5)
7. Evolve MAN from APE (5)
8. Prove a ROGUE to be a BEAST (10)
9. Pay COSTS in PENCE (9)
10. Raise ONE to TWO (7)
11. Change BLUE to PINK (8)
12. Change BLACK to WHITE (6)
13. Turn WITCH into FAIRY (12)
14. Make WINTER SUMMER (13)
15. Make BREAD into TOAST (6)

If you can solve any of the Doublets with fewer linking words, more power to you.

Carroll invented riddles for his child friends. At the mad tea party in *Alice in Wonderland,* the mad hatter asks, "Why is a raven like a writing-desk?" but he doesn't provide the answer. Many able paradoxologists have struggled with the riddle. Their answers are included in Martin Gardner's *The Annotated Alice.* "In keeping with Carroll's alliterative style," Gardner writes, "[Sam] Loyd offers as his best solution: because the notes for which they are noted are not noted for being musical notes."

In the magazine *The Rectory Umbrella* Carroll published his most famous conundrum, "Difficulty No. 2." "Which is the best, a clock that is right only once a year, or a clock that is right twice every day? 'The latter,' you reply, 'unquestionably.' Very good, reader, now attend.

"I have two clocks: one doesn't go *at all,* and the other loses a minute a day: which would you prefer? 'The losing one,' you answer, 'without a doubt.' Now observe: the one which loses a minute a day has to lose twelve hours, or seven hundred and twenty minutes before it is right again, consequently it is only right once in two years, whereas the other is evidently right as often as the time it points to comes around, which happens twice a day. So you've contradicted yourself *once.* 'Ah, but,' you say, 'what's the use of its being right twice a day, if I can't tell when the time comes?' Why, suppose the clock points to eight o'clock, don't you see that the clock is right *at* eight o'clock? Consequently when eight o'clock comes your clock is right. 'Yes, I see *that*,' you reply. Very good, then you've contradicted yourself *twice:* Now get out of the difficulty as you can, and don't contradict yourself again if you can help it."

Carroll's poems and letters are riddled with conundrums, puns, acrostics, and other forms of diabolical wordplay. He often formed anagrams of people's names. In the Dormouse's story in *Alice in Wonderland,* there is a girl named Lacie, whose name of course is an anagram of Alice. Carroll once came up with three anagrams of one name: "Wild agitator! Means well," "A wild man will go at trees," and "Wilt tear down all images?" Carroll's other anagrams include "Ah! We dread an ugly knave," "Flit on, cheering angel," and "I lead, Sir."

Mirror images, backward writing, and other examples of the looking-glass motif can be found in much of Carroll's writing. He coined the term *semordnilap (palindromes* in reverse) to describe a word that is another word spelled backward. He was tickled by the god-dog reversal. In *Sylvie and Bruno,* Carroll's evangelistic fantasy, Bruno notes that *evil* is *live* spelled backward. In a letter to a child named Enid, Carroll wrote "Suppose we read *backwards,* what does Mr. Dodgson like best to do?" The answer: "He likes best to go to DINE."

Carroll appropriated the term *portmanteau* to describe a word (for example, *slithy)* formed from the sounds and meanings of two or more other words. A portmanteau is a suitcase that opens into two compartments. Humpty Dumpty tells Alice that the words in "Jabberwocky" are "like a portmanteau—there are two meanings packed up into one word." For example, *frumious* is a combination of *furious* and *fuming, galumphing* of *gallop* and *triumphant,* and *mimsy* of *miserable* and *flimsy.* The snark in *The Hunting of the Snark* is apparently a snaillike shark (or, according to other interpretations, a snakelike shark or a creature that snarls and barks). In *Sylvie and Bruno* there is the delightful *y'reince*, "The Unpronounceable Monosyllable," which is a phonetic contraction of *your Royal Highness*. In a letter to a friend Carroll described the "Jabberwocky" word *uffish* as "a state of mind when the voice is gruffish, the manner roughish, and the temper huffish." It is fun to try to decompose the portmanteaus in "Jabberwocky" that Carroll did not explain. Any guesses about *chortle* and *vorpal*? Eric Partridge thinks that *chortle* comes from *chuckle* and *snort*, and *vorpal* from *voracious* and *narwhal*.

Carroll's lighthearted wordplay rubbed off on Vladimir Nabokov, who translated *Alice in Wonderland* into Russian in 1922. There are *semordnilaps* in Nabokov's *Pale Fire*. "In an old barn. She twisted words: pot, top,/ Spider, redips. And 'powder' was 'red wop.'" Nabokov also notes that "T. S. Eliot" in reverse is "toilest." In the "Commentary" that follows *Pale Fire* Nabokov mentions Doublets: "My illustrious friend showed a childish predilection for all sorts of word games and especially for so-called word golf. He would interrupt the flow of a prismatic conversation to indulge in this particular pastime, and naturally it would have been boorish of me to refuse playing with him. Some of my records are: hate-love in three, lass-male in four, and live-dead in five (with 'lend' in the middle)."

Like Carroll, Nabokov was fond of anagrams. In *Lolita* Humbert Humbert learns from *Who's Who in the Limelight* about the playwright Vivian Darkbloom, whose name is an anagram of Vladimir Nabokov. (Before Dodgson settled on the pen name Lewis Carroll, he considered the pseudonyms Edgar Cuthwellis and Edgar U. C. Westhall, which are anagrams of his Christian names, Charles and Lutwidge.) In *Ada,* the story of a ninety-year romance between a brother and a sister, Nabokov often uses the words *nicest* and *insect*, which are anagrams of *incest.*

Nabokov picked up Carroll's passion for portmanteaus. *Lolita* includes *libidream* (from *libido* and *dream*), *mauvemail* (a weak version of blackmail), *honeymonsoon* ("They were going to India for their honeymonsoon"), *earwitness* ("a child shamming sleep to earwitness primal sonorities") and *Purpills* (from *Papa's Purple Pills,* with which Humbert tries to drug Lolita so that he can possess her).

Both Carroll and Nabokov were as fond of taking words apart

THERAPIST

as they were of putting them together. Carroll called attention to the *love* in *glove* and the *ink* in *wink*. Nabokov wondered about the significance of the *rapist* in *therapist*, the *ass* in *passion*, and the *jest* in *majesty*.

PASSION

Little girls tickled the fancies of both Lewis Carroll and Vladimir Nabokov. All of Nabokov's works are bursting with literary allusions. In *Lolita,* which is the confession of Humbert Humbert, who craves little girls, there are references to *Alice in Wonderland.* "A breeze from wonderland" affects Humbert's thoughts and, later, Humbert envisages "a half-naked nymphet stilled in the act of combing her Alice-in-Wonderland hair." Nevertheless, Carroll is never mentioned directly, which is quite odd considering that Nabokov once said—according to Alfred Appel, Jr.—"I always call him Lewis Carroll Carroll because he was the first Humbert Humbert." In an interview in *Wisconsin Studies* Nabokov explained the omission of Carroll's name in *Lolita:* "He has a pathetic affinity with H.H. but some odd scruple prevented me from alluding in *Lolita* to his wretched perversion and to those ambiguous photographs he took in dim rooms. He got away with it, as many other Victorians got away with pederasty and nympholepsy. His were sad scrawny little nymphets, bedraggled and half-undressed, or rather semi-undraped, as if participating in some dusty and dreadful charade."

GIRL =
PHALLUS?

Alice in Wonderland, written by a man who was ferociously fond of little girls, begins with a long fall down a rabbit hole. Should it come as a surprise that the book and its author have been subjected to Freudian psychoanalysis? One analyst suggests that Alice is the embodiment of the symbolic equation Girl = Phallus, which "is most strikingly, and almost undisguisedly, presented in reality in the form, figure, and function of the tambour majorette, a young girl marching in front of a large body of men, united with them but still demonstratively put in front of them. . . . Supposedly, she is loved by the men of the unit. . . . After her parade she is not treated like a woman, but like a tired little child, badly in need of rest. She then behaves like a penis post coitum. She emphasizes her symbolic meaning by her high hat, preferably adorned by a feather or pompon."

More than one analyst has claimed—in a leap of logic worthy of Evel Knievel—that Carroll's fascination with inverted images suggests he was an invert. The Jungians have also chimed in, because "as Jung has stated, the man possessed by the *anima* sees all of life as a game or puzzle."

From his grave Carroll could not respond to his analysts. Nabokov, for his part, led a lifelong crusade against "King Sigmund" and "the Viennese delegation." In *Pale Fire* Nabokov pokes fun at Freudian analysis by quoting two passages from the psychoanalytic literature. Dr. Oscar Pfister wrote in *The Psychoanalytic Method:* "By picking the nose in spite of all commands to the contrary, or when a youth is all the time sticking his finger through his buttonhole . . . the analytic teacher knows that the appetite of the lustful one knows no limit in his phantasies." Erich Fromm concluded in *The Forgotten Language* that "The little cap of red velvet in the German version of Little Red Riding Hood is a symbol of menstruation."

In many ways *Lolita* is a parody of a psychoanalytic case study, and it is unrelenting in lampooning the "Viennese healers." When Humbert recuperated at a sanatorium from a bout with depression, he discovered "an endless source of robust enjoyment in trifling with psychiatrists: cunningly leading them on . . . teasing them with fake 'primal scenes'; and never allowing them the slightest glimpse of one's real sexual predicament. By bribing a nurse I won access to some files and discovered, with glee, cards calling me 'potentially homosexual' and 'totally impotent.'"

In "the cryptogrammic paper chase," Humbert discovers in motel registers the mocking, pseudonymous signatures of the fiend who stole his Lolita. One of the pseudonyms is Dr. Kitzler, Eryx, Miss; "and any good Freudian, with a German name and some interest in religious prostitution, should recognize at a glance the implication of 'Dr. Kitzler, Eryx, Miss.'" In *Keys to Lolita* Carl Proffer deciphers the doctor's name. *Kitzler* is German for clitoris and Eryx is a mountain in Sicily on which there is a temple to Venus. "The priestesses were prostitutes, i.e, Eryx Misses," Proffer explains. "The Mount of Venus also comes to mind."

In interviews, as well as in his fiction, Nabokov kept up his assault on psychoanalysis. In *The Annotated Lolita* Alfred Appel, Jr., quotes from two interviews. On National Educational Television Nabokov said of Freud: "I think he's crude, I think he's medieval, and I don't want an elderly gentleman from Vienna with an umbrella inflicting his dreams upon me. *I* don't have the dreams that he discusses in his books. I don't see umbrellas in my dreams. Or balloons." To Appel Nabokov said: "Let the credulous and the vulgar continue to believe that all mental woes can be cured by a daily application of old Greek myths to their private parts. I really do not care."

Carroll, unlike Nabokov, did some serious work in philosophy, although his philosophizing is couched in wacky tales. In 1895 he published a charming story, "What The Tortoise Said to Achilles," in the quarterly *Mind.* The story, which is as philosophically insightful as it is tongue-in-cheek, takes place after Achilles, the star of Zeno's paradoxes, "had overtaken the Tortoise, and had seated himself comfortably on its back."

The Tortoise then asked Achilles if he admired Euclid's seminal work on the foundation of geometry. "Passionately!" Achilles replied. "So far, at least, as one *can* admire a treatise that won't be published for some centuries to come!"

The Tortoise summarized one of Euclid's first arguments as follows, where A and B are the premises and Z is the conclusion:

A. Things that are equal to the same are equal to each other.
B. The two sides of this Triangle are equal to the same.
Z. The two sides of this Triangle are equal to each other.

"Readers of Euclid will grant, I suppose," said the Tortoise, "that Z follows logically from A and B, so that anyone who accepts A and B as true, must accept Z as true?"

"Undoubtedly!" replied Achilles. "The youngest child in a High School—as soon as High Schools are invented, which will not be till some two thousand years later—will grant *that*."

The Tortoise then got to the crux of the matter. If a person accepts A and B as true and accepts no other premises, is he then compelled out of logical necessity to accept that Z is true? The answer is no, because the argument entails an additional premise, call it C: If A and B are true, Z must be true.

In other words, the whole argument can be reiterated as follows:

A. Things that are equal to the same are equal to each other.
B. The two sides of this Triangle are equal to the same.
C. If A and B are true, Z must be true.
Z. The two sides of this Triangle are equal to each other.

Achilles was satisfied. "If you accept A and B and C," he said, "you must accept Z."

"And why *must* I?" asked the Tortoise.

"Because it follows *logically* from them. If A and B and C are true, Z *must* be true. You don't dispute that, I imagine?"

"If A and B and C are true, Z must be true," the Tortoise thoughtfully repeated. "That's *another* Hypothetical, isn't it? And, if I failed to see its truth, I might accept A and B and C, and *still* not accept Z, mightn't I?"

Achilles conceded the point, and introduced another premise, called D: If A and B and C are true, Z must be true. As you now undoubtedly must realize, the Tortoise was still not satisfied and insisted that Achilles explicitly recognize another premise (If A and B and C and D are true, Z must be true) and so on ad infinitum.

Carroll noted that the Tortoise was last overheard saying, "Have you got that last step down? Unless I've lost count, that makes a thousand and one. There are several millions more to come. And *would* you mind as a personal favour, considering what a lot of instruction this colloquy of ours will provide for the Logicians of the Nineteenth Century?"

ANSWERS

Here are Carroll's solutions to the Doublets.

1. PIG
   WIG
   WAG
   WAY
   SAY
   STY

2. WHEAT
   CHEAT
   CHEAP
   CHEEP
   CREEP
   CREED
   BREED
   BREAD

3. PEN
   E'EN
   EEL
   ELL
   ILL
   ILK
   INK

4. TEARS
   SEARS
   STARS
   STARE
   STALE
   STILE
   SMILE

5. EYE
   DYE
   DIE
   DID
   LID

6. POOR
   BOOR
   BOOK
   ROOK
   ROCK
   RICK
   RICH

7. APE
   ARE
   ERE
   ERR
   EAR
   MAR
   MAN

8. ROGUE
   VOGUE
   VAGUE
   VALUE
   VALVE
   HALVE
   HELVE
   HEAVE
   LEAVE
   LEASE
   LEAST
   BEAST

9. COSTS
   POSTS
   PESTS
   TESTS
   TENTS
   TENTH
   TENCH
   TEACH
   PEACH
   PEACE
   PENCE

10. ONE
    OWE
    EWE
    EYE
    DYE
    DOE
    TOE
    TOO
    TWO

11. BLUE
    GLUE
    GLUT
    GOUT
    POUT
    PORT
    PART
    PANT
    PINT
    PINK

12. BLACK
    BLANK
    BLINK
    CLINK
    CHINK
    CHINE
    WHINE
    WHITE

13. WITCH
    WINCH
    WENCH
    TENCH
    TENTH
    TENTS
    TINTS
    TILTS
    TILLS
    FILLS
    FALLS
    FAILS
    FAIRS
    FAIRY

14. WINTER
    WINNER
    WANNER
    WANDER
    WARDER
    HARDER
    HARPER
    HAMPER
    DAMPER
    DAMPED
    DAMMED
    DIMMED
    DIMMER
    SIMMER
    SUMMER

15. BREAD
    BREAK
    BLEAK
    BLEAT
    BLEST
    BLAST
    BOAST
    TOAST

"Wild agitator! Means well," "A wild man will go at trees," and "Wilt tear down all images?" are anagrams of William Ewart Gladstone. "Ah! We dread an ugly knave" is an anagram of Edward Vaughan Kennealy, "Flit on, cheering angel" is an anagram of Florence Nightingale, and "I lead, Sir!" is an anagram of Disraeli.

CHAPTER 17

The Talk of the Town

IF YOU WERE PACKING your personal effects for a long night in the air-raid shelter, you'd certainly include the Deadly Double, Chicago's favorite game of dice and chips. That's the thrust of the bizarre advertisement reproduced here from the November 22, 1941, issue of *The New Yorker*. "WARNING!" the ad cries out in German, English, and French. The section that shows the dice appeared fourteen times in the issue; the section that discusses air-raid-shelter preparations appeared only once. Hidden in the ad is a shocking message that sends chills up even my back. Lora passed out when she discovered the meaning of the concealed message. Can you find and decrypt it? I have deciphered it on page 180.

Achtung
WARNING
alerte !

See Advertisement Page 70

MONARCH PUBLISHING CO.
New York

Achtung
WARNING
Alerte!

We hope you'll never have to spend a long winter's night in an air-raid shelter, but we were just thinking . . . it's only common sense to be prepared. If you're not too busy between now and Christmas, why not sit down and plan a list of the things you'll want to have on hand. . . . Canned goods, of course, and candles, Sterno, bottled water, sugar, coffee or tea, brandy, and plenty of cigarettes, sweaters and blankets, books or magazines, vitamin capsules . . . and though it's no time, really, to be thinking of what's fashionable, we bet that most of your friends will remember to include those intriguing dice and chips which make Chicago's favorite game

THE
DEADLY DOUBLE

$2.50 at leading Sporting Goods and Department Stores Everywhere

ANSWER

The eerie advertisement forecast the bombing of Pearl Harbor on December 7, 1941. All six faces of the dice (XX, 12, 24, 5, 0, and 7) contribute information about the time and place of the attack. The 12 represents the month of the attack and the 7 represents the day.

The place to be attacked is indicated by rough coordinates. The 24 is the latitude. The longitude is obtained from the XX, the 5, and the 0. Because X is both a multiplication sign and the Roman numeral for 10, it may stand for 10 multiplied by 10, or 100. One hundred plus 50 (a combination of the 5 and the 0) gives a longitude of 150. Pearl Harbor is the closest significant spot to the junction of these coordinates if the 150 is longitude west of Greenwich. Of course, if the longitude is east, Guam would appear likely. However, it is possible that the use of old German script for the *A* in *Achtung* (the old German *A* is shaped like a *W*) may indicate west longitude.

There is another possible interpretation. The XX may stand for the double cross the Japanese committed. The 0 means "zero hour" and 5 stands for the approximate hour at which the bombers took off. (The attack itself occurred about 8:00 A.M. Pearl Harbor time.)

Why was the advertisement used to forecast the attack? The most likely explanation is that the Japanese devised it as a means of warning their intelligence agents in the United States of the time and place. They undoubtedly worked out the warning technique months earlier, and so their men must have been diligently reading *The New Yorker* from week to week.

The true story behind the ad may never emerge because of the untimely death of the man who placed it. The advertisement was placed by a Roger Paul Craig, who owned the Monarch Publishing Company mentioned in the ad. According to an article in *The Investigator,* Craig's widow claims that he died at the age of thirty-six on May 26, 1946. After dining with friends in a New York restaurant, Craig slipped into a sidewalk delivery chute and never came out alive. (Could the three sixes—the sign of the devil—in Craig's age and date of death also be a code?)